WETLAND
AND
WATER
RESOURCE
MODELING
AND
ASSESSMENT

A Watershed Perspective

Integrative Studies in Water Management and Land Development

Series Editor
Robert L. France

Published Titles

**Boreal Shield Watersheds: Lake Trout Ecosystems
in a Changing Environment**
Edited by J.M. Gunn, R.J. Steedman, and R.A. Ryder

The Economics of Groundwater Remediation and Protection
Paul E. Hardisty and Ece Özdemiroğlu

**Forests at the Wildland–Urban Interface:
Conservation and Management**
Edited by Susan W. Vince, Mary L. Duryea, Edward A. Macie,
and L. Annie Hermansen

Handbook of Water Sensitive Planning and Design
Edited by Robert L. France

Porous Pavements
Bruce K. Ferguson

Restoration of Boreal and Temperate Forests
Edited by John A. Stanturf and Palle Madsen

**Wetland and Water Resource Modeling and Assessment:
A Watershed Perspective**
Edited by Wei Ji

WETLAND
AND
WATER
RESOURCE
MODELING
AND
ASSESSMENT

A Watershed Perspective

Edited by Wei Ji

CRC Press
Taylor & Francis Group
Boca Raton London New York

CRC Press is an imprint of the
Taylor & Francis Group, an **informa** business

CRC Press
Taylor & Francis Group
6000 Broken Sound Parkway NW, Suite 300
Boca Raton, FL 33487-2742

First issued in paperback 2019

© 2008 by Taylor & Francis Group, LLC
CRC Press is an imprint of Taylor & Francis Group, an Informa business

No claim to original U.S. Government works

ISBN-13: 978-1-4200-6414-8 (hbk)
ISBN-13: 978-0-367-38800-3 (pbk)

Library of Congress Cataloging-in-Publication Data

Wetland and water resource modeling and assessment : a watershed perspective
/ editor, Wei Ji.
    p. cm. -- (Integrative studies in water management and land development)
    ISBN 978-1-4200-6414-8 (alk. paper)
    1. Wetlands--Remote sensing. 2. Wetland ecology. 3. Ecological mapping.
4. Water resources development. 5. Watershed management. 6. Geographic
information systems. I. Ji, Wei, 1955- II. Title. III. Series.

GB622.W48 2007
553.7--dc22   2007024968

Visit the Taylor & Francis Web site at
http://www.taylorandfrancis.com

and the CRC Press Web site at
http://www.crcpress.com

Contents

PART I Geospatial Technologies for Wetland Mapping

PART II   Wetland Hydrology and Water Budget

Chapter 6   Study on the Intra-Annual Distribution Characteristics of the
                 Water Budget in the Hilly Region of Red Soil in Northeast

                 Junfeng Dai, Jiazhou Chen, Yuanlai Cui, and Yuanqiu He

Chapter 7   Forest and Water Relations: Hydrologic Implications

                 *Ge Sun, Guoyi Zhou, Zhiqiang Zhang, Xiaohua Wei,
                 Steven G. McNulty, and James Vose*

PART III Water Quality and Biogeochemical Processes

PART IV   Wetland Biology and Ecology

PART V   Watershed Assessment and Management

Chapter 20 Integrated Modeling of the Muskegon River: Tools for
Ecological Risk Assessment in a Great Lakes Watershed 247

Michael J. Wiley, Bryan C. Pijanowski, R. Jan Stevenson,
Paul Seelbach, Paul Richards, Catherine M. Riseng,
David W. Hyndman, and John K. Koches

Chapter 21 Watershed Management Practices for Nonpoint Source
Pollution Control 259

Shaw L. Yu, Xiaoyue Zhen, and Richard L. Stanford

Foreword:
A Wider View of Wetlands

There are few landforms that have been treated with the same degree of distrust, distaste, disdain, and destruction as have wetlands (e.g., R. France, ed. 2007. *Wetlands of Mass Destruction: Ancient Presage for Contemporary Ecocide in Southern Iraq*). Part of the reason for this comes about from the reality that we both literally and figuratively do not quite know where we stand in relation to wetlands; they are neither land, nor water, but exist as some uncomfortable nether region situated between the two (R. France, ed. 2008. *Healing Natures, Repairing Relationships: New Perspectives on Restoring Ecological Spaces and Consciousness*). Our language also reflects the pejorative view that much of society has of wetlands, for who has not felt "swamped" at one time or another by being "bogged" down through having too much work due to being "mired" in details?

The present volume, edited by Wei "Wayne" Ji, offers a counterpoint to such a gloomy worldview. Wetlands as described in these pages are shown to be very much centers of hydrological and ecological importance in the landscape, a view that would have certainly found resonance with that nineteenth-century wetland enthusiast, Henry David Thoreau (R. France. 2003. *Profitably Soaked: Thoreau's Engagement with Water*). And it is here, with its overall message of demonstrating the cardinal need to reinsert wetlands back into their landscape, where the present book succeeds most admirably. Wetlands are not isolated entities but rather influence, and are in turn influenced by, a vast variety of environmental and anthropogenic factors (R. France. 2003. *Wetland Design: Principles and Practices for Landscape Architects and Land-Use Planners*). In order to preserve the environmental integrity of wetlands it is necessary to circumvent the strange imbalance that exists between the scale at which wetland losses are felt by society and the scale at which wetlands have traditionally been studied or managed. The present book, *Wetland and Water Resource Modeling and Assessment: A Watershed Perspective*, anchors the goal of holistic management in a firm scientific grounding.

The bulk of the chapters in the present volume originated from a conference. Editor W. Ji took the wise step, however, as taken in several other volumes in this series (R. France, ed. 2002. *Handbook of Water Sensitive Planning and Design*; and R. France. 2008. *Handbook of Regenerative Landscape Design*), to actively solicit contributions from others who did not present at the conference in order to better address the book's overall objective. The result is a well-rounded whole with the myriad subjects being truly catholic in scope, including, for example, various spatial mapping approaches, hydrological models, ecological appraisals, and water quality and biogeochemistry investigations, many directed toward understanding threats on wetlands posed by climate change and water imbalances, chemical contamination and eutrophication, and land-use alterations and soil erosion, to name just a few. In this respect, the present book is a worthy addition to the aspirations of the series Integrative Studies in Water Management and Land Development by Taylor & Francis. In the end, it is only by recognizing the essentialness of a watershed approach for understanding and managing landscapes (e.g., R. France, ed. 2005. *Facilitating Watershed Management: Fostering Awareness and Stewardship*; R. France. 2006. *Introduction to Watershed Development: Understanding and Managing the Impacts of Sprawl*) that wetlands can be properly assessed and modeled, as the collective voices of the authors reiterate time and again within these pages.

Finally, the international scope of the present volume is worth noting — an additional attribute of this series whose previous books have featured case studies from North America, Southeast Asia, and much of Europe, in addition to Australia and Brazil. The majority of this volume's authors originate from China, also the location of much of the research contained herein. Perhaps this should not be surprising. The Chinese have long recognized the multifaceted importance of wetlands. For example, I begin chapter 1, "Foundations," in my primer *Wetland Design: Principles and Practices for Landscape Architects and Land-Use Planners* (W.W. Norton, 2003) with the following sentence: "Early in the last millennium, a Chinese military commander retired to the old picturesque town of Suzhou. There, by drawing water from one of its famous canals, he created a marvelous garden retreat for emotional and spiritual peace . . . [the] *Chanlang ting* (Pavilion of Blue Waves)." The present book continues this foundational tradition, in this case emphasizing the importance of viewing wetlands in a watershed perspective.

Robert L. France
Harvard University

Preface:
Toward a Watershed Perspective

This is a book about the methods and geospatial techniques for modeling and assessing wetlands and water resources at the watershed scale. As background, I would like to start with a brief introduction with an example from Poyang Lake. Situated in Jiangxi Province, it is the largest freshwater lake in China, with many marshes, grasslands, and alluvial floodplains in its watershed. The wetland area of Poyang Lake Basin has diverse flora and fauna and provides important habitats for many migratory birds. As a wetland of international importance (referred to as "Poyanghu" on the Ramsar List of Wetlands of International Importance, 1992) with a unique land use history, Poyang Lake has attracted great attention, domestically and internationally, from research and conservation organizations. In 2004 the Chinese Ministry of Education established a facility for lake and watershed research—the Key Lab of Poyang Lake Ecological Environment and Resource Development, which is housed in Jiangxi Normal University. The lab soon became very instrumental in attracting scientists and scholars for collaborative research. Between June 27 and 30, 2005, the lab organized and hosted a productive academic meeting at Jiangxi Normal University: The International Conference on Poyang Lake Complex Environment System and Advanced Workshop on Watershed Modeling and Water Resources Management. This event attracted scholars and professionals from China, North America, and Europe, who presented research findings and technical developments related to issues in wetland and water resource science and management.

Many of the papers presented at that conference are included in this book. However, the book is not simply the conference proceedings. The editorial advisory board selected the conference papers and also invited papers from recognized experts in order to better present the theme of the book. All submissions were peer reviewed and the best of them appear in this volume.

The theme of the book, wetland and water resource modeling and assessment, is an active field of research that constantly undergoes theoretical and technical innovations. This book emphasizes a watershed perspective in the modeling and assessment areas. The term *watershed* means a geographic area where water drains into a body of water such as a river, lake, or wetland. Other terms are often used to describe the same concept, such as *river basin*, *drainage basin*, and *catchment area*. Since the early 1990s, *watershed management* or the *watershed approach*—coordinated resource planning and management based on hydrologically defined geographic areas—has been promoted as a common strategy of water resource stewardship and other related environmental activities. In 2000 the U.S. Departments of Agriculture and the Interior announced a unified policy to protect water quality and aquatic ecosystems on federal lands. The policy serves as a framework for land and resource management focused on watersheds. This policy has been supported by the U.S. Departments of Commerce, Defense, and Energy; the Environmental Protection Agency; the Tennessee Valley Authority; and the U.S. Army Corps of Engineers. Similar trends have occurred on other continents. For example, a pilot study on integrated water management, launched by NATO/CCMS (North Atlantic Treaty Organization/Committee on the Challenges of Modern Society) in 2002, was conducted through a series of workshops involving representatives from the NATO countries. While I was traveling in the Biebrza National (wetlands) Park in Poland in the summer of 2006, I had the good fortune to participate in part of this pilot study's sixth workshop held there. I was impressed by the vision and breadth of the study for implementing water management based on the "river basins" across Europe. Within Jiangxi Province, China, the surface water of approximately 95% of its land drains into Poyang Lake through several major rivers. The hydrological feature of this watershed is quite unique and important to water resources, biology, ecology, and socioeconomic development in the region. Thus, the watershed perspective in the research and management of the Poyang Lake ecosystem has a long tradition.

Sound watershed-based water resource planning and management should rest on scientifically justifiable data and innovative technical tools. Thus, assessment and modeling of key processes of terrestrial and aquatic ecosystems are crucial to the success of watershed management, which is becoming, as demonstrated in the studies included in this book, an active field of research and technical development. With a watershed perspective, ecosystem assessment and modeling commonly possess the following major characteristics: (1) Sufficiently large spatial scales in data collection and analysis in order to encompass major watershed features. This often leads to using remote sensing and GIS (geographic information system) for data acquisition and integration, as well as for spatial analysis. (2) Inclusion of landscape features in order to appropriately characterize watershed hydrological processes and related ecosystem components. This usually requires relating land cover and land use dynamics to water features. (3) Linking assessment or modeling results to management decisions for specific objectives. This often results in the development of decision support tools to facilitate ecosystem assessment under various management scenarios and criteria. These characteristics of watershed assessment and modeling can be found in many of the studies included in this book.

The book is divided into five parts. Part I focuses on geospatial methods and technologies. It includes four research projects on improving remote sensing methods for wetland mapping, which has comprised a fundamental yet challenging area of study for detecting wetlands at a watershed level. The chapters in this part cover topics ranging from expert system techniques for improving the remote sensing identification of wetlands (Torbick et al.; Cai and Chen), to the use of hyperspectral imagery in identifying salt marshes (Yang et al.), to remote sensing spectral techniques for vegetation mapping (Chen et al.).

Part II concentrates on wetland hydrology and water budget. McNulty et al. use a modeling framework to assess the interannual water supply stress over the next 40 years across the southern United States as a function of climate, groundwater supply, and population change. Focusing on the red-soil hilly region of Poyang Lake basin, the work of Dai et al. illustrates the characteristics of temporal distribution of a water budget, which helps us understand the occurrence of seasonal droughts and to adopt better measures to increase water use efficiency. The chapter by Sun et al. is based on a synthesis of existing worldwide literature on the relations between forestation and watershed hydrology. It identifies the factors affecting hydrologic responses to forestation and discusses the potential hydrologic consequences of large-scale, vegetation-based watershed restoration efforts in China. Carried out in the Xing Feng Catchment within the Zhujiang Watershed, the study by Wen et al. introduces the use of a modified TOPMODEL to simulate streamflow and distinguish subsurface stormflow from the baseflow. The chapter contributed by Croley and He provides a description of the development and application of a spatially distributed, physically based surface/subsurface model of hydrology and water quality, which is used to evaluate both agricultural nonpoint-source and point-source pollution loadings at the watershed level.

Part III addresses issues relating to water quality and biogeochemical processes at the watershed scale. In their chapter, He and Croley introduce the application of the model introduced in Part II in the Cass River Watershed, a subwatershed of the Saginaw Bay watersheds in the Great Lakes area, to estimate the potential of nonpoint-source pollution loadings. The chapter by Gui et al. demonstrates the use of SWAT, an existing watershed assessment model, to simulate changes of nutrients at a temporal scale of one hundred years in Honghu Lake Basin, China. The work of Varnakovida et al. describes the construction of a model to predict total nitrogen, total phosphorous, and total suspended solid concentrations in lakes based on surrounding land cover and land use types and patterns.

Part IV is devoted to issues of wetland biology and ecology. The chapter by Li et al. introduces a method of predicting annual soil losses in Xiushui Watershed in Jiangxi Province using integrated data concerning precipitation, topography, soil, and vegetation cover with GIS. Lougheed et al. describe an investigation that develops and tests field-based methods for the rapid assessment of wetland conditions in Muskegon River Watershed, Michigan. In their chapter, Guo and Chen introduce a geospatial techniques–based method of deriving appropriate indicators for analyzing ecological conditions in Poyang Lake Watershed. Focusing on the feeding habitat of the endangered Siberian crane wintering at Poyang Lake, the chapter by Wu et al. propose a conceptual framework for integrating a model of plant biomass with remote sensing and GIS methods to simulate the growth and biomass of one submerged aquatic species under various hydrological conditions. Qi et al. present new research that expands traditional remote sensing to acoustic sensing. Their goal is to improve our knowledge about the usefulness of acoustic signals as a means to measure and interpret ecological characteristics of a landscape—the soundscape.

Part V features innovative development and applications of wetland assessment and management methodologies. The chapter by Ji and Ma covers the research, development, and application issues concerning geospatial decision models for assessing wetland vulnerability to human impact at a watershed scale. They provide prewarning information for regulatory wetland management decision making. Using the study of Muskegon River Watershed in the Great Lakes Region as an example, Stevenson et al. contribute a chapter on a conceptual framework of watershed science. It comprehensively reviews and discusses watershed science as related to its essential role in watershed management, its complex nature and the solutions for complex watershed problems, and its implementation in a multidisciplinary and collaborative framework. Also dealing with Muskegon River Watershed, Wiley et al. demonstrate the development of a GIS-based approach that uses ecologically defined valley segment units to integrate a landscape transformation model with a variety of hydrologic and other models for assessing risks to key watershed resources under various scenarios. The major thrust of the chapter by Yu et al. is a discussion of recent advances in watershed management technology for nonpoint source pollution control. It also discusses a number of issues that should be addressed before implementing watershed pollution-source control measures.

The contributors include senior scholars and young researchers. All of the chapters were peer reviewed. Hayley Charney at Michigan State University helped edit some of the chapters. As the editor of the book, I thank all of the reviewers for their time and dedicated work, which made this book infinitely better. I want to recognize the crucial role of the Key Lab of Poyang Lake Ecological Environment and Resource Development of the Chinese Ministry of Education at Jiangxi Normal University in organizing the international conference and the workshop in 2005, that contributed many chapters to the book. I am pleased that this book has been selected for the Integrative Studies in Water Management and Land Development book series, and am honored that Dr. Robert France at Harvard University, the series editor, has written the foreword for it.

Wei "Wayne" Ji

Editor

Dr. Wei "Wayne" Ji is a professor of geosciences at the University of Missouri, Kansas City (UMKC). He has taught courses in geographic information systems (GIS), remote sensing, biogeography, and landscape ecology at UMKC since 1999.

Over the past 25 years, his research has focused primarily on the study of water environments using geospatial methods. At Peking University, China, he completed his master's thesis focusing on remote sensing of water quality. His PhD dissertation research at the University of Connecticut developed a new model for coastal bathymetry with satellite remote sensing as well as a GIS for coastal mapping. In the 1990s he conducted research at the National Wetlands Research Center of the U.S. Geological Survey, where he studied innovative geospatial methods for wetland ecosystem restoration and management in Louisiana and south Florida. During that period he developed a decision support GIS for wetland value assessment modeling for coastal wetland restoration planning, and a prototype decision support GIS for wetland permit assessment. Dr. Ji also proposed a decision modeling method for integrating the results of computer simulations of wildlife species for evaluating effects of different wetland restoration scenarios.

With the support of the U.S. Environmental Protection Agency, his recent research focused on geospatial decision models for assessing wetland vulnerability to potential human impacts, for application to urban wetland studies. In addition to the wetland issues in the United States, Ji has a long-term interest in the Poyang Lake Watershed in China—a wetland area of international importance. With a U.S. Fulbright senior scholar award for research in Germany, in 2006 he surveyed wetlands and collected related research information in the coastal areas of Germany, Poland, and the Netherlands in order to understand the impact of the historical east–west division of that region on coastal resources, especially wetlands. In addition to wetlands, Ji also studied long-term landscape effects of urban sprawl in metropolitan Kansas City, GIS-based methods for assessing the conservation status of wildlife genetic diversity through a case study in the southern Appalachians, and spatial distributions of wintering birds in the lower Mississippi region.

Dr. Ji has served as a manuscript reviewer for many academic journals and a proposal reviewer for agencies like NASA and the U.S. National Science Foundation. He was the guest editor for *Marine Geodesy*'s 2003 special issue on marine and coastal GIS. He is an associate editor of *Wetlands* — an international journal published by the Society of Wetland Scientists.

Editorial Advisory Board

Contributors

Francisco J. Artigas
Meadowlands Environmental Research
  Institute
Lyndhurst, New Jersey

Jeb Barzen
World Center for the Study and
  Preservation of Cranes
Baraboo, Wisconsin

Elly P.H. Best
Environmental Laboratory
U.S. Army Corps of Engineers
  Research and Development Center
Vicksburg, Mississippi

Pearl Bonnell
Michigan Lakes and Streams
  Association
Long Lake, Michigan

S. Biswas
Department of Electrical and Computer
  Engineering
Michigan State University
East Lansing, Michigan

James Burnham
World Center for the Study and
  Preservation of Cranes
Baraboo, Wisconsin

Thomas M. Burton
Center for Water Sciences
Department of Zoology
Michigan State University
East Lansing, Michigan

Xiaobin Cai
State Key Laboratory of Information
  Engineering in Surveying,
  Mapping and Remote Sensing
Wuhan University
Wuhan, China

Jiazhou Chen
College of Resource and
  Enviroment
Huazhong Agricultural University
Wuhan, China

Liangfu Chen
State Key Laboratory of Remote
  Sensing Science
Institute of Remote Sensing
  Applications
Chinese Academy of Sciences
Beijing, China

Shuisen Chen
Guangzhou Institute of Geography
Guangzhou, China

Xi Chen
State Key Laboratory of Hydrology —
  Water Resources and Hydraulic
  Engineering
Hohai University
Nanjing, China

Xiaoling Chen
State Key Laboratory of Information
  Engineering in Surveying, Mapping
  and Remote Sensing
Wuhan University
Wuhan, China

Yongqin Chen
Chinese University of Hong Kong
Hong Kong, China

Erika C. Cohen
Southern Global Change Program
United States Department of
  Agriculture Forest Service
Raleigh, North Carolina

Thomas E. Croley II
Great Lakes Environmental Research
  Laboratory
National Oceanic and Atmospheric
  Administration
Ann Arbor, Michigan

Yuanlai Cui
State Key Laboratory of Water
  Resources and Hydropower
  Engineering Science
Wuhan University
Wuhan, China

Kevin Czajkowski
Department of Geography and
  Planning
University of Toledo
Toledo, Ohio

Junfeng Dai
State Key Laboratory of Water
  Resources and Hydropower
  Engineering Science
Wuhan University
Wuhan, China

Stuart H. Gage
Computational Ecology and
  Visualization Laboratory
Department of Entomology
Michigan State University
East Lansing, Michigan

Feng Gui
Nanjing Institute of Geography and
  Limnology
Chinese Academy of Sciences
Nanjing, China

Peng Guo
State Key Laboratory of Information
  Engineering in Surveying,
  Mapping and Remote Sensing
Wuhan University
Wuhan, China

Chansheng He
Department of Geography
Western Michigan University
Kalamazoo, Michigan

Yuanqiu He
Institute of Soil Science
Chinese Academy of Sciences
Nanjing, China

R. Anton Hough
Department of Biology
Wayne State University
Detroit, Michigan

David W. Hyndman
Department of Geological Sciences
Michigan State University
East Lansing, Michigan

Wei "Wayne" Ji
Department of Geosciences
University of Missouri
Kansas City, Missouri

Weitao Ji
Bureau of Jiangxi Poyang Lake
  National Nature Reserve
Nanchang, China

Wooyeong Joo
Computational Ecology and
  Visualization Laboratory
Michigan State University
East Lansing, Michigan

John K. Koches
Annis Water Institute
Grand Valley State University
Muskegon, Michigan

Geying Lai
Nanjing Institute of Geography and
  Limnology
Chinese Academy of Sciences
Nanjing, China

Patrick Lawrence
Department of Geography and
  Planning
University of Toledo
Toledo, Ohio

Jan de Leeuw
International Institute for Geo-
  Information Science and Earth
  Observation
Enschede, Netherlands

Hui Li
State Key Laboratory of Information
  Engineering in Surveying, Mapping
  and Remote Sensing
Wuhan University
Wuhan, China

Jian Li
The Key Lab of Poyang Lake
  Ecological Environment and Resource
  Development
Jiangxi Normal University
Nanchang, China

Qinhuo Liu
State Key Laboratory of Remote
  Sensing Science
Institute of Remote Sensing
  Applications
Chinese Academy of Sciences
Beijing, China

Yaolin Liu
School of Resource and Environmental
  Sciences
Wuhan University
Wuhan, China

David T. Long
Department of Geological Sciences
Michigan State University
East Lansing, Michigan

Vanessa L. Lougheed
Department of Biological Sciences
University of Texas at El Paso
El Paso, Texas

Jia Ma
Department of Geosciences
University of Missouri
Kansas City, Missouri

Steven G. McNulty
Southern Global Change Program
United States Department of
  Agriculture Forest Service
Raleigh, North Carolina

Joseph P. Messina
Center for Global Change and Earth
  Observations
Department of Geography
Michigan State University
East Lansing, Michigan

Jennifer A. Moore Myers
Southern Global Change Program
United States Department of
  Agriculture Forest Service
Raleigh, North Carolina

Brian Napoletano
Department of Forestry and Natural
  Resources
Purdue University
West Layfayette, Indiana

Christian A. Parker
Department of Zoology
Michigan State University
East Lansing, Michigan

Bryan C. Pijanowski
Department of Forestry and Natural
  Resources
Purdue University
West Lafayette, Indiana

Jiaguo Qi
Center for Global Change and Earth
  Observations
Department of Geography
Michigan State University
East Lansing, Michigan

Paul Richards
Department of Biology
State University of New York
Brockport, New York

Catherine M. Riseng
School of Natural Resources
University of Michigan
Ann Arbor, Michigan

Paul Seelbach
Institute for Fisheries Research
Michigan Department of Natural
  Resources
Ann Arbor, Michigan

Richard L. Stanford
Department of Civil
  Engineering
University of Virginia
Charlottesville, Virginia

Alan D. Steinman
Annis Water Institute
Grand Valley State University
Muskegon, Michigan

R. Jan Stevenson
Center for Water Sciences
Department of Zoology
Michigan State University
East Lansing, Michigan

Xiaobo Su
The Key Lab of Poyang Lake
  Ecological Environment and Resource
  Development
Jiangxi Normal University
Nanchang, China

Ge Sun
Southern Global Change Program
United States Department of
  Agriculture Forest Service
Raleigh, North Carolina

Liqiao Tian
State Key Laboratory of Information
  Engineering in Surveying, Mapping
  and Remote Sensing
Wuhan University
Wuhan, China

Nathan Torbick
Department of Geography
Michigan State University
East Lansing, Michigan

Donald G. Uzarski
Annis Water Institute
Grand Valley State University
Muskegon, Michigan

Pariwate Varnakovida
Center for Global Change and Earth
  Observations
Department of Geography
Michigan State University
East Lansing, Michigan

Valentijn Venus
International Institute for
  Geo-Information Science and
  Earth Observation
Enschede, Netherlands

James Vose
Coweeta Hydrologic Laboratory
United States Department of
  Agriculture Forest Service
Otto, North Carolina

Yeqiao Wang
Department of Natural Resources
  Science
University of Rhode Island
Kingston, Rhode Island

Xiaohua Wei
University of British Columbia
Kelowna, Canada

Pei Wen
State Key Laboratory of Hydrology –
  Water Resources and Hydraulic
  Engineering
Hohai University
Nanjing, China

Narumon Wiangwang
Center for Global Change and Earth
  Observations
Department of Geography
Michigan State University
East Lansing, Michigan

Michael J. Wiley
School of Natural Resources
University of Michigan
Ann Arbor, Michigan

Guofeng Wu
International Institute for
  Geo-Information Science and
  Earth Observation
Enschede, Netherlands

Zhongyi Wu
State Key Laboratory of Information
  Engineering in Surveying,
  Mapping and Remote Sensing
Wuhan University
Wuhan, China

Jiansheng Yang
Department of Natural Resources
  Science
University of Rhode Island
Kingston, Rhode Island

Ge Yu
Nanjing Institute of Geography and
  Limnology
Chinese Academy of Sciences
Nanjing, China

Shaw L. Yu
Department of Civil
  Engineering
University of Virginia
Charlottesville, Virginia

Zhiqiang Zhang
Beijing Forestry University
Beijing, China

Xiaoyue Zhen
Tetra Tech, Inc.
Fairfax, Virginia

Guoyi Zhou
Southern China Botany Garden
Chinese Academy of Sciences
Guangzhou, China

Part I

Geospatial Technologies for Wetland Mapping

1 Application and Assessment of a GIScience Model for Jurisdictional Wetlands Identification in Northwestern Ohio

Nathan Torbick, Patrick Lawrence,
and Kevin Czajkowski

1.1 INTRODUCTION

Wetlands are natural ecosystems subject to permanent or periodic inundation or prolonged soil saturation sufficient for the establishment of hydrophytes and/or the development of hydric soils or substrates unless the environmental conditions are such that they prevent them from forming (Cowardin et al. 1979, Tiner 1999). Wetlands provide a range of environmental and socioeconomic benefits. These benefits range from their ability to store floodwaters and improve water quality, to providing habitat for wildlife and supporting biodiversity, to aesthetic values (Mitsch and Gos selink 2000). The loss of wetlands has gained considerable attention over the past few decades. New policies and regulations require improved wetlands management practices and information to promote wetland ecosystems.

The utilization of satellite remote sensing technology for inventorying and identifying wetlands has proven to be a useful and frequent application (i.e., Hardinsky et al. 1986, Kindscher et al. 1998, Lunetta and Balogh 1999, Townsend and Walsh 2001). Remote sensing techniques are often less costly and time-consuming for large geographic areas compared to conventional field mapping methods. Satellite data provides regular overpass intervals that enable the monitoring of wetland changes seasonally and over longer time periods. Nearly every sensor has been tested and utilized for wetlands identification and wetlands-related research (Ozesmi and Bauer 2002). Each sensor has advantages and limitations often related to their associated resolutions: spatial, temporal, radiometric, spectral.

Multitemporal imagery is often utilized to overcome resolution limitations by incorporating phenological information. Johnston and Barson (1993) used

3

three-season imagery multispectral thematic mapper (TM) data, however they report wide-ranging accuracies and state that remote sensing imagery is unlikely to replace aerial photography methodologies for wetlands mapping. Lunetta and Balogh (1999) used multispectral Landsat 5 TM to identify jurisdictional wetlands and compared single-season imagery with multitemporal imagery. Using aerial photography and field-collected data, classification accuracies range from 69% for single-season to 88% for two-season imagery. Townsend and Walsh (2001) used three-season imagery to achieve accuracies above 90% when classifying TM data for forested wetlands.

A variety of classification techniques have been executed using multispectral data. These range from visual interpretation to expert systems. The term *expert system* is a general descriptor for a variety of organizational frameworks such as intelligent systems, artificial neural networks, or knowledge-based systems. Here, expert system refers to a knowledge database, or wetlands rule-based model, incorporating geospatial data. In the past decade expert systems have contributed to improving wetlands remote sensing science, often by addressing the resolution limitations. Using satellite imagery together with other data sources can improve wetlands classification detail. Ozesmi and Bauer (2002) synthesize the literature indicating that generally multitemporal imagery and ancillary information, such as soils, elevation, or other maps, improve wetlands classification. Bolstad et al. (1992), Sader et al. (1995), and Lunetta and Balogh (1999) found that using wetlands-related ancillary data, including field data collections, can significantly improve classifications.

In this study, an expert system was developed integrating multiple types of data to identify jurisdictional wetland types of interest in northwestern Ohio. A secondary goal of the project was to develop a systematic methodology that does not require ambiguous human visual interpretation methodology. With increases in availability of multispectral imagery and digital geospatial data, developing consistent methods for identifying wetlands is advantageous. This study incorporated agricultural land use, land cover, and soils data to improve wetlands multispectral remote sensing capabilities. A GIS rule-based decision tree algorithm was designed to classify four primary wetland types of interest: forest, prairie, riparian, and coastal wetlands. The expert system was very effective in depicting wetland types. However, the complexity of jurisdictional wetlands regulations and the model have limitations in the application and ease of use.

1.2 STUDY AREA

In the past 200 years, over 90% of Ohio's wetlands have been destroyed or altered as a result of human impacts (Dahl 2000, Great Lakes Commission 2002). The Great Lakes Basin watershed was once dominated by a variety of wetland ecosystems. This watershed is now an intensely cultivated area and a patched network of shrinking wetlands. The study site is Lucas County, Ohio, within the Maumee River and Lake Erie watersheds in northwestern Ohio (Figure 1.1). Lucas County, with a population of approximately one-half million, contains a mix of land use and land covers including agriculture, industry, residential, forest, and urban systems. The city of Toledo, with a population of approximately 330,000, is located at the mouth of the Maumee River, which flows into Maumee Bay of the western basin of Lake Erie.

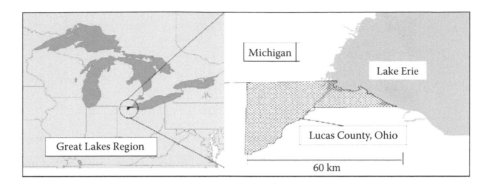

FIGURE 1.1   Lucas County is located within the Maumee River watershed and drains into Lake Erie.

The Maumee River has the largest drainage area of any Great Lakes river (total river basin covering 21,538 km^2) with 6,344 stream km draining into the Maumee River. Approximately 85% of the total river drainage area is considered to be agriculture (Maumee RAP 2004).

1.3 METHODS

1.3.1 GEOSPATIAL DATA

Landsat 7 Enhanced Thematic Mapper Plus (ETM+) imagery was used as the remotely sensed data. ETM+ imagery from 2001 through 2003 was examined for phenological cycle changes, cloud cover, and overall image quality. In processing the satellite imagery, subsets of the study region were performed on all scenes to reduce computational requirements. Scenes from the early spring (March 14, 2001) leaf-off condition and the late summer (August 21, 2001) peak growth condition were selected for the study area. Each wetland type can receive different inputs for identification, including imagery requirements. The specific image inputs are discussed in model development and classification.

Three main geospatial ancillary layers were utilized. The first was a database containing information on agricultural land use for the study region. The data is an extensive spatial descriptor of agricultural practices and types for the study region. The data was originally developed by county-level government for real estate tax information. Farmers and related agricultural land users have land registered by associated land use and land cover information where land owners pay taxes on current agriculture use instead of its development potential (Auditors Real Estate Information System [AREIS] 2003). Approximately 83% of the agricultural land identified in the database is considered as tillable and being tilled. Tillable land parcels make up roughly 30% of the county (Figure 1.2).

The second dataset was soils information for the entire study region supplied by the U.S. Department of Agriculture (USDA) Soil Conservation Service (SCS), now part of the Natural Resources Conservation Service (NRCS). NRCS has an extensive digital taxonomy system to identify soil characteristics and soil types such as hydric

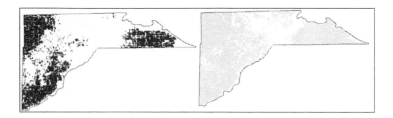

FIGURE 1.2   Left: parcels identified as tillable in black. Right: hydric soils shown in gray.

soil condition parameters. Using NRCS guidelines, hydric soils were extracted into a single hydric soils area of interest (AOI) layer. Approximately 50% of the study region, or 41,671 hectares, was considered hydric. Combining the agricultural and soils datasets, a specialized AOI was created by eliminating any agricultural parcel identified as tilled from the hydric soil AOI.

The third data source was a stream network map. A hydrology network layer was created from digitized U.S. Geologic Survey topographic maps combined with digitized county-level drainage patterns. Buffer zones were created at 60 and 90 meters as a riparian zone for the county.

1.4 WETLANDS FIELDWORK AND REGULATIONS

Training sites were developed for the four wetlands types of interest; forest, prairie, coastal, and riparian. Training site development included extensive field visits and assessments. Wetland delineations were carried out at each training site. Delineations and assessments performed at each site included global positioning systems (GPS) collection, soil core logs examining hydric indicators, soil taxonomy cross-validation with the NRCS digital database, vegetation transects and strata sampling, hydrologic and surface hydrology examination, and digital picture catalogs.

Hydrological requirements for the study region for jurisdictional wetlands include inundation or soil saturation, usually within 12 inches of the surface, for more than 5% of the growing season. If assessing for wetlands in drier months this quickly becomes problematic. Other indicators for wetlands hydrology include watermarks, drift lines, sediment deposits, and drainage patterns. Secondary indicators include abundant oxidized rhizospheres within 12 inches of the surface, water-stained leaves, wetland vegetation, and soils survey soil-water information. The seasonality changes and hydrology indicators are an illustration of the complexity of wetlands classification. Large efforts need to be put into assessing wetlands classifications rather than appearance of surface water. Hydrology indicators at each training site included prolonged inundation or soil saturation for extensive periods during the growing season.

Soil log samples were cross-referenced using the Munsell® guide for determining soil properties. Soil chromas, hues, values, and moisture levels were examined. Other soil characteristics such as redoxamorphic features, mottles, contrasts, textures, and general descriptors confirmed hydric qualifiers (USDA 2002).

Vegetation transect-sampled plots examined plant species and plant community structure to provide information on wetland vegetation indicator status. Plants were cataloged as ranging between obligate plant status, occurring with an estimated probability >99% of the time in wetland conditions, to facultative plant status, occurring with an estimated probability between 33% and 67% of the time in wetland conditions, to upland plant status, occurring with an estimated probability <1% of the time in wetland conditions. Most training site vegetation indicators were categorized as obligate wetland plants or facultative wetland plants.

1.4.1 CLASSIFICATION

A decision tree classification algorithm was created to classify the satellite data as the main component of the expert system. Decision trees have substantial advantages for remote sensing classification problems because of their flexibility, intuitive simplicity, and computational efficiency (Defries et al. 1998, Hansen et al. 1996, Friedl and Brodley 1997). Decision trees predict class membership by recursively partitioning data into more homogenous subsets (Defries and Chan 2000). Used effectively in many studies, decision tree algorithms have not been used extensively for higher-end spatial resolution data, that is, Landsat (Brown de Colstoun et al. 2003). In the expert system, a hierarchical decision tree hypothesis (wetland class of interest), was created with rules defining the integrated geospatial data for the wetlands classification model. Here, the wetlands class of interest and rules are developed using the soils and agricultural spatial layers defining limitations on possible classifications.

After training data collection and selection, the decision tree with four wetland class signatures was developed. Pixel values for each band were examined and extracted from the wetland class training sites to develop class signatures. Due to the fact that wetlands types are an isolated, patched landscape within the study site, a limited amount of pixels could be examined to formulate the class signatures. One advantage to using the decision tree method is the acceptability of nonparametric data (Foody et al. 1995, Friedl et al. 1999, Brown de Colstoun et al. 2003). This allows for specific classes to be identified based on nonparametric signatures developed from training site objects rather than statistical parameters. Figure 1.3 displays the conceptual framework of the decision tree. This example shows that a possible ETM+ pixel response is given an allowable range for each band. For each band, all the pixels within the training sites had pixel values extracted to develop the possible range of spectral responses.

The different wetland class types required different inputs into the decision tree model. All wetland classes used the constraining soils and agricultural layers as rules defining classifiable areas. The geospatial AOIs limited possible classifications

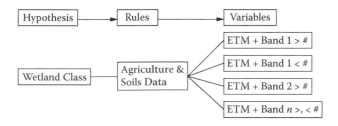

FIGURE 1.3 Conceptual framework of wetlands expert system.

to nontillable, nonagriculture land with hydric soils. For coastal wetlands, a two-kilometer (km) buffer zone, as defined by the Great Lakes Commission (2002), was applied to the coastline of Lake Erie. This new buffer shapefile was imported into the expert system model as a rule to constrain possible coastal wetland classifications to within 2 km of the shoreline. The coastal wetlands only used a single-season late spring leaf-off image from mid-March. Prairie wetlands, forest wetlands, and riparian wetlands utilized a two-season (leaf-off and peak phenological cycle) multitemporal stacked image from mid-March and late August.

1.4.2 ASSESSMENT

An error matrix (Congalton and Green 1999) was constructed to evaluate expert system performance. The expert system framework was designed only to classify wetland types of interest, rather than classify all land covers in the study area. Therefore misclassifications could not be placed into other mutually exclusive categories. For this reason the *other* category was created to locate misclassifications. Overall accuracy, omission and commission patterns, and classification performance were examined with the error matrix.

A stratified random sampling scheme with 50 points per class was determined as an adequate (Congalton and Green 1999) sampling size for the validation. The accuracy points were checked using a variety of techniques. Initially all points were displayed over the satellite imagery and model outputs. Approximately half of the points could be determined from black-and-white digital orthorectified aerial photographs. The remaining points that could not be determined from the aerial photographs were ground truthed in the field. Field ground truthing during the late spring and early summer months under wetland delineation guidelines (Tiner 1993) of the remaining locations provided an increased database detailing vegetation communities, site descriptions, GPS locations, digital pictures, soil types, and associated wetlands variables at each point. This was done between May and August (spring and summer) of 2003.

1.5 RESULTS

The first simulation that was assessed had an extremely high overall accuracy value. Overall accuracy is a measure of the total number of correct sample points divided by the total number of classification sample points. The overall classification accuracy was 96.5%. Out of 144 points, 139 points were correctly classified. Six total points were thrown out as not being accessible. However, this high overall accuracy value is somewhat misleading.

Qualitatively, many errors of omission existed on iteration one of the expert system. The constraints on the allowable band values were decreased by 15% to include more pixels in an attempt to classify more pixels to reduce false negatives. This decrease was chosen after numerous adjustments followed by assessments, which led to a final tradeoff at 15%. This resulted in iteration two of the expert system. Table 1.1 is an error matrix for the second iteration with reduced model restrictions. The total number of pixels classified, or wetlands type identified, nearly doubled for each category. The overall accuracy was 84%. Wetland prairie omission errors decreased

TABLE 1.1

Error matrix for expert system iteration two.

Class Name	Other	Coastal	Forest	Prairie	Column Totals	Producer's Accuracy (%)	User's Accuracy (%)
Other	0	0	0	0	0	0	0
Coastal Wetland	2	48	0	0	50	98	96
Forest Wetland	2	1	47	0	50	77	94
Prairie Wetland	3	0	14	22	39	100	56
Row Totals	7	49	61	22	139		

substantially as well. However, the user's accuracy for wetland prairie decreased to 56% with many misclassifications of which the majority included scrub/shrub-type cover identified.

1.6 DISCUSSION

The assessment process was time consuming and often more difficult than a simple correct/incorrect classification interpretation when trying to identify jurisdictional wetlands. Examining wetland characteristics and performing field wetland delineations was an involved process. The stratified random points often fell on private property, remote locations, or areas difficult to place into a mutually exclusive category. This was particularly the case with wetland prairie misclassifications as scrub/shrub land cover.

The extremely high overall accuracy of iteration one is a reflection of the model and its complexity. The number of pixels classified and the relative accuracies between the expert system iterations is another indication of model performance. The number of classified pixels for wetland types nearly doubled for iteration two; inversely, the accuracy for wetland prairies decreased to approximately only half being correct.

Two main reasons, both related to the complexity of wetlands classification generally, contributed to misclassifications. First is the pattern of wetlands. In the study region, as in many locations, wetlands are often isolated systems creating a spotty network across the landscape. Even with the 30-meter spatial resolution of Landsat bands, subpixel landscape heterogeneity was frequently moderate. A given sample point, or wetland classified pixel, was a mixture of vegetation covers that pushed the limits of ETM+ spatial resolutions to accurately classify the area even with the ancillary spatial data.

A second reason for misclassifications related to the complexity of wetlands was the expert system classes of interest. The model developed extrapolates spectral response data by identifying desired pixels within the model rules. This was required with soil moisture and hydrology playing such a large role in wetlands. A tremendous amount of in-class spectral variability occurs within each wetland type, as well as spectral overlap among the different types of wetlands. The variation among the wetland spectral signatures developed in the expert system from

ETM+ bands 1 through 7 was moderate. Test iterations were executed without using the soils and agriculture-constraining AOIs, and the number of misclassified pixels increased beyond an acceptable level. Experiment iterations increasing the range of allowable spectral responses were tested in order to decrease the pixilation. This also increased the number of misclassified pixels.

Thus, in order to allow for accurate classification without incorporating misclassified pixels, additional classes of interest are required. If high levels of omission error are present and large amounts of spectral variation exist, additional classes are needed. For example, a forest wetland class can have many plant species. As well, forested wetlands can occur on a number of soil types and all have variations in hydrology. In the general region, twenty-nine wetland plant communities were identified (Anderson 1982). When spectral restrictions were decreased, pixels from other land covers were classified incorrectly. Adding additional wetland class categories captured those classes without sacrificing accuracy. For example, some prairie wetland pixels were not classified. Adding a scrub/shrub wetland class would likely have increased the number of wetland pixels classified without removing pixels identified as prairie wetland. This was the case for prairie wetlands as reflected in the assessment values between the two iterations described. Determining the ecological or regulatory cutoff point between classes was challenging even with the wetland delineation procedures.

1.7 CONCLUSION

The objective of this paper was to develop a wetlands classification system to identify the dwindling wetland land cover types in Lucas County in northwestern Ohio. Much debate exists with respect to defining wetlands, particularly regarding agricultural wetlands and human disturbance of the three primary wetland indicators. In this case, the expert system contributed considerably to supplying relatively accurate and detailed maps for the county using current jurisdictional regulations. The system outlined showed that multispectral Landsat image data congruent with ancillary data can provide increased classification information. A general compromise between omission and class level accuracy was required for the project.

ACKNOWLEDGMENTS

This material is based in part upon work supported by Natural Areas Stewardship, Inc. and the Ohio Environmental Protection Agency 319 Grant Program. Satellite imagery was provided by the OhioView Remote Sensing Consortium and the Ohio Library and Information Network (OhioLINK). This work is adapted from Torbick et al., 2006, JEMREST.

REFERENCES

Anderson, D. 1982. *Plant communities of Ohio: A preliminary classification and description.* Columbus: Ohio Department of Natural Resources, Division of Natural Areas and Preserves.

Auditors Real Estate Information System (AREIS). Lucas County, Ohio Information Services. http://www.co.lucas.oh.us/AREIS/areismain.asp.

Boldstad, P., M. Wehde, and R. Linder. 1992. Rule-based classification models: Flexible integration of satellite imagery and thematic spatial data. *Photogrammetric Engineering and Remote Sensing* 58: 965–971.

Brown de Colstoun, E., M. Story, C. Thompson, K. Commisso, T. Smith, and J. Irons. 2003. National park vegetation mapping using multitemporal Landsat 7 data and a decision tree classifier. *Remote Sensing of Environment* 85:316–327.

Congalton, R., and K. Green. 1999. *Assessing the accuracy of remotely sensed data: Principles and practices.* New York: Lewis Publishers.

Cowardin, L., V. Carter, F. Golet, and E. LaRoe. 1979. *Classification of wetlands and deepwater habitats of the United States.* U.S. Fish and Wildlife Service. Biological services program. FWS/OBS-79/31.

Dahl, T. 2000. *Status and trends of wetlands in the conterminous United States 1986 to 1987.* Washington, DC: U.S. Department of the Interior. Fish and Wildlife Service.

Defries, R., and J. Chan. 2000. Multiple criteria for evaluating machine learning algorithms for land cover classification from satellite data. *Remote Sensing of Environment* 74:503–515.

Defries, R., M. Hansen, J. Townsend, and R. Sohlberg. 1998. Global land cover classification at 8 km spatial resolution: The use of training data derived from Landsat imagery in decision tree classifiers. *International Journal of Remote Sensing* 19:3141–3168.

Foody, G., M. McCulloch, and W. Yates. 1995. The effect of training set size and composition on neural network classification. *International Journal of Remote Sensing* 16:1707–1723.

Friedl, M., and C. Brodley. 1997. Decision tree classification of land cover from remotely sensed data. *Remote Sensing of Environment* 61:399–409.

Friedl, M., C. Brodley, and A. Strahler. 1999. Maximizing land cover classification accuracies produced by decision trees at continental to global scales. *IEEE Transactions on Geoscience and Remote Sensing* 37:969–977.

Great Lakes Commission (GLC). 2002. *An overview of U.S. Great Lakes areas of concern.* Great Lakes National Program. U.S. Environmental Protection Agency Report Series. Washington, DC: U.S. Environmental Protection Agency.

Hansen, M., R. Dubayah, and R. Defries. 1996. Classification trees: An alternative to traditional land cover classifiers. *International Journal of Remote Sensing* 17:1075–1081.

Hardisky, M., M. Gross, and V. Klemas. 1986. Remote sensing of coastal wetlands. *BioScience* 36:453–460.

Ji, W., J. Johnston, M. McNiff, and L. Mitchell. 1992. Knowledge-based GIS: An expert system approach for managing wetlands. *Geo Info Systems* 10:60–64.

Johnston, R., and M. Barson. 1993. Remote sensing of Australian wetland: An evaluation of Landsat TM data for inventory and classification. *Austrailian Journal of Marine Freshwater Resources,* 44:235–252.

Kindscher, K., A. Fraser, M. Jakubauskas, and D. Debinski. 1998. Identifying wetland meadows in Grand Teton National Park using remote sensing and average wetland values. *Wetlands Ecology and Management* 5:265–273.

Lunetta, R., and M. Balogh. 1999. Application of multitemporal Landsat 5 TM imagery for wetland identification. *Photogrammetric Engineering & Remote Sensing* 65:1303–1310.

Maumee RAP. 2004. Maumee Remedial Action Plan. http://www.maumeerap.org/.

Mitsch, W., and J. Gosselink. 2000. *Wetlands.* New York: John Wiley and Sons.

Moran, M., R. Jackson, P. Slater, and P. Teillet. 1992. Evaluation of simplified procedures for retrieval of land surface reflectance factors from satellite sensor output. *Remote Sensing of Environment* 41:169–184.

Ozesmi, S., and M. Bauer. 2002. Satellite remote sensing of wetlands. *Wetlands Ecology and Management* 10:381–402.

Sadar, S., D. Ahl, and W. Liou. 1995. Accuracy of Landsat-TM and GIS rule based methods for forest wetland classification in Maine. *Remote Sensing of Environment* 53:133–144.

Tiner, R. 1993. The primary indicators method: A practical approach to wetland recognition and delineation in the United States. *Wetlands* 13:50–64.

Tiner, R. 1999. *Wetland indicators: A guide to wetland identification, classification, and mapping.* Boca Raton, FL: Lewis Publishers.

Townsend, P., and S. Walsh. 2001. Remote sensing of forested wetlands: Application of multitemporal and multispectral satellite imagery to determine plant community composition and structure in southeastern USA. *Plant Ecology* 157:129–149.

U.S. Department of Agriculture, Soil Conservation Service. 1980. *Soil survey of Lucas County.* Washington, DC: U.S. Department of Agriculture.

U.S. Department of Agriculture, National Resources Conservation Service. 2002. *Field indicators of hydric soils in the United States, Version 5.0.* Ed. G. W. Hurt, P. M. Whited, and R. F. Pringle. Fort Worth, TX: USDA, NRCS in cooperation with the National Technical Committee for Hydric Soils.

2 An Expert System–Based Image Classification for Identifying Wetland-Related Land Cover Types

Xiaobin Cai and Xiaoling Chen

2.1 INTRODUCTION

Understanding wetland changes in relation to urban development using remote sensing techniques is critical to planning ecosystem management and sustainable regional development. The per-pixel classification methods have been used extensively in thematic information extraction (Wang et al. 2004). The spectral property of surface objects was utilized in remote sensing classification in these techniques. However, the spectral property information usually is not sufficient to generate accurate classification results. Therefore, other types of information, such as DEM (digital elevation model) and road data, have been used to improve classification accuracy.

The intrinsic spatial distribution and relationships that exist in geographical objects enable the adoption of spatial information, such as the texture of an object, which was widely used in the process of classification at the pixel level. On the other hand, to better represent spatial relationships, spatial information at the object level was more useful in identifying the distribution patterns of different classes. This idea directly resulted in the invention of object-oriented classification software. Ecognition, a software package, was recently enhanced by integrating the object-oriented classification techniques with its classification procedures. However, Ecognition's classification procedure is too complicated (Burnett et al. 2003), especially when selecting segmentation parameters. This paper introduces a simple object-oriented method to improve classification accuracy.

2.2 STUDY AREA AND DATA

The study area is located in northern Jiangxi Province, China, with a geographic range of E115° 24′ to 117° 43′ and N27° 57′ to 29° 47′. The area covers Poyang Lake and its surrounding wetlands, which is one of 21 wetlands in China that have been designated as having international importance under the Ramsar Convention.

FIGURE 2.1   The edited SRTM DEM in the study area.

Lushan National Park, one of UNESCO's world heritage sites, is also located in the study area.

The image selected is the Landsat ETM+ image taken on October 28, 2004. The Shuttle Radar Terrain Mission (SRTM)–edited DEM was used as the complementary data. The SRTM DEM data has the highest resolution among datasets with a near-global (i.e., between S 56° and N 60°) coverage. It has sufficiently detailed topographic information and can be derived to fit the selected image classification. Figure 2.1 shows the edited SRTM DEM in the study area. As the altitude changes from high to low, a color gradient from blue to red was used in the legend to represent corresponding altitude differences. The central part of the image was dominated by Poyang Lake Plain, shown as blue in the DEM.

2.3 METHODOLOGY AND RESULTS

Normally it is quite difficult to acquire satisfactory classification accuracy because of the spectral similarity of different objects. By examining spectral characteristics, three groups of objects with spectral similarity were identified: (a) wetland and grassland/forest, (b) bare land and developing urban area, and (c) built-up area and muddy beach. Figure 2.2 illustrates the special overlap between the built-up area and the muddy beach.

To improve the classification accuracy, the road data and DEM (the geographic information system [GIS] data) were incorporated in the classification process. The maximum likelihood classification was used to identify initial boundaries of objects, with the water bodies being extracted and masked from the image. In this process, the spatial relationships among the classified objects and the GIS data were referenced to develop decision rules in the expert system in order to better distinguish the spectrally mixed objects.

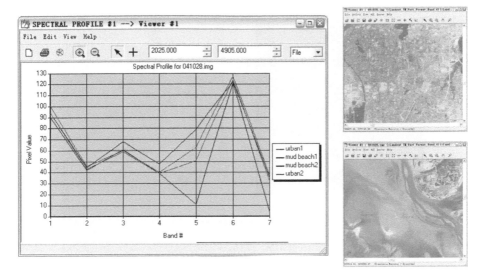

FIGURE 2.2 The spectral comparison of the built-up urban area with the muddy beach.

2.3.1 Water Body Identification and Supervised Classification

Generally, water bodies can be easily identified using the Normalized Difference Water Index (NDWI) as follows (McFeeters 1996, Gao 1996):

$$NDWI = (GREEN - NIR)/(GREEN + NIR) \tag{2.1}$$

where GREEN is the brightness value of the green band and NIR is the brightness value of the near-infrared band. However, it is still difficult to use the NDWI (Normalized Difference Water Index) to separate the surface waters with plants from the plants in "wet" lands. To address this problem, the Normalized Difference Vegetation Index (NDVI), a good indicator of plants, was also calculated.

$$NDVI = (NIR - RED)/(NIR + RED) \tag{2.2}$$

where RED is the brightness value of the red band. Then the difference between NDVI and NDWI was used to detect water bodies more accurately.

Supervised classification was then employed to extract other relevant cover information from the image by masking out the water bodies. Seven land cover types, including built-up area, wetland (with vegetation), grassland, forest, bare land, muddy beach, and farmland were identified.

2.3.2 Expert Knowledge

According to spectral analysis of these land covers, the muddy beach is easily confused with the built-up area, while muddy beaches usually distribute along water

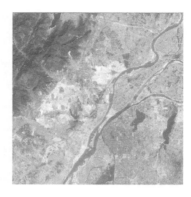

FIGURE 2.3 The identified developing urban area (in the left image, the lighter area is the developing urban area identified; the right image is the Landsat ETM+ RGB image with Bands 4, 3, and 2). **(See color insert after p. 162.)**

bodies and built-up areas are close to roads and rails. Thus, the spatial adjacency of these two objects to water bodies, the province-level roads, and railways were measured on the relevant thematic map, and were then incorporated in the expert system. Spectrally, the wetland (with vegetation) is similar to the forest and the grassland. However, wetlands are usually located adjacent to waters or muddy beaches, with their elevations lower than surrounding forests and grasslands. The normal water level of Poyang Lake fluctuates between 5.9 m and 22.20 m seasonally and the height of the hygrophyte inhabiting the area ranges from 13 m to 16 m (Tan 2002). Therefore, the neighborhood relationship and the wetland classification unit's mean elevation were used as criteria in the expert system. As the spectral property of the developing urban area is similar to bare land, it was not classified as an individual land cover type in the initial supervised classification. Usually the developing urban area has more stable relationships with roads and existing urban areas than bare land; this fact was used as the decision rule in the classification to distinguish the two types of classes. Figure 2.3 shows the classified results. In the left picture the red part is the developing areas while the existing urban areas and roads were depicted as blue or green lines. The result accords well with the color features on the original ETM+ image.

2.3.3 OBJECT NEIGHBORHOOD SEARCH AND DEM ANALYSIS

To acquire the spatial neighborhood relationship at an object level, a clump procedure was performed on the image with ERDAS Imagine. The results included an object ID image, where the pixel value stands for its object ID code. A search was applied to the reference object images, which included a water object image and a road object image at the pixel level, to produce a new neighborhood relationship image. Then the neighborhood ID image was obtained with an overlay of the object ID image and the neighborhood relationship image, where the LUT (look up table) was achieved. A new LUT, including only the pixels with a value larger than zero, was produced by sieving the LUT using the histogram of the neighborhood ID image. The new LUT was loaded into the object ID image to produce a neighborhood object image. With the expert knowledge analysis described in section 2.3.2, five object-level search images were created through object neighborhood searching. To obtain the altitude property of each classified object, the overlay analysis was utilized between the clump result of the supervised classification and the edited SRTM DEM data, which generated the mean elevation of each classified object.

2.3.4 CONSTRUCTION OF THE EXPERT SYSTEM

The expert system was constructed with the ERDAS Imagine's Knowledge Engineering function, where the prospective class was used as a hypothesis, the determination criterion as a rule, and the parameter as a variable. In the experiment, the supervised classification result, the five object-level search images, and the mean elevation in every object clump were induced as the variables. The criteria were expressed as the following rules:

If (muddy-ser = T and (sc = g or sc = fr) and men-elv ≤ 20) finalc = wt;

If (muddy-ser = T and sc = u and (rod-ser = T or ral-ser = T or urb-ser = T) and wat-ser = T) finalc = m;

If (muddy-ser = T and sc = u and (rod-ser = T or ral-ser = T or urb-ser = T) and wat-ser = T) finalc = fm;

If (sc = bl and (rod-ser = T or ral-ser = T or urb-ser = T)) finalc = dv;

Else finalc = sc.

where *muddy-ser* denotes the muddy beach object-searching image, *rod-ser* the provincial road object-searching image, and *ral-ser* the railway object-searching image. *T* represents the object that was located in the searching threshold; the *finalc* means the final expert system classification result, and *sc* is the supervised classification result. Grassland is denoted by *g*, forest *fr*, wetland *wt*, muddy beach *m*, built-up urban areas *u*, farmland *fm*, bare land *bl* and developing urban area *dv*.

A classified result based on the expert system was obtained, and the comparison between the expert system classification result and the supervised classification results are shown in Figures 2.4 and 2.5. The pictures illustrate that some errors generated by the per-pixel classification were corrected by the expert system–based classification. For example, the previously misclassifed built-up areas were corrected to the muddy beaches and the previously missing developing urban areas near Nanchang City were detected by the new method.

2.4 CONCLUSION

The rule-based expert system approach could improve the classification of wetland-related objects that have similar spectral characteristics. The object-level spatial searching method, which was incorporated with supervised classification results and the road and elevation information, proved to be effective for deriving a set of object spatial searching images.

Water
Built-up Urban
Welland
Mud Beach
Forest
Grassland
Bare Land
Farmland
Developing Urban Area

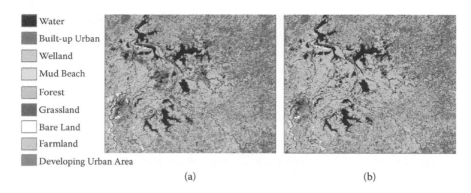

(a)   (b)

FIGURE 2.4   The overall comparison between the per-pixel classification (a) and the expert system–based classification (b). **(See color insert after p. 162.)**

FIGURE 2.5   The comparison for the Poyang Lake area with the two classification methods (the left image is the per-pixel classification result, the middle one the original Landsat ETM+ image, and the right the expert system–based classification result). **(See color insert after p. 162.)**

ACKNOWLEDGMENTS

This study was funded by the 973 Program (2003CB415205) of the Chinese Natural Science Foundation and the Open Fund project (200401006(1) of the Key Lab of Poyang Lake Ecological Environment and Resource Development of Chinese Ministry of Education housed in Jiangxi Normal University.

REFERENCES

Burnett, Charles, Kiira Aaviksoo, and Stefan Lang. 2003. An object-based methodology of mapping mires using high resolution imagery. Paper presented at Ecohydrological Processes in Northern Wetlands Tallinn. Estonia: Tartu University Press.

Gao, B. 1996. NDWI—a normalized difference water index for remote sensing of vegetation liquid water from space. *Remote Sensing of Environment* 58:257–266.

McFeeters, S. K. 1996. The use of the Normalized Difference Water Index (NDWI) in the delineation of open water features. *International Journal of Remote Sensing* 17:1425–1432.

Tan, Qulin. 2002. Study on remote sensing change detection and its application to Poyang international importance wetland. Ph.D. dissertation, Institute of Remote Sensing Applications, Chinese Academy of Sciences, 48–49.

Wang, Zivu, Wenxia Wei, Shuhe Zhao, and Xiuwan Chen. 2004. Object-oriented classification and application in land use classification using SPOT-5 PAN imagery. Geoscience and Remote Sensing Symposium, *IGARSS '04 Proceedings*, Vol. 5, 3158–3160.

3 Mapping Salt Marsh Vegetation Using Hyperspectral Imagery

Jiansheng Yang, Francisco J. Artigas, and Yeqiao Wang

3.1 INTRODUCTION

Salt marshes are the transition between submerged and emerged environments and are among the most biologically productive ecosystems in the world. Not only do salt marshes experience a variety of physical characteristics, they also offer significant ecological benefits (Ko and Day 2004). Salt marshes provide habitats for a wide variety of fish and wildlife, and help maintain coastal water quality by acting as filters and scrubbers of sediments and excess nutrients (Herrera-Silveira et al. 2004).

Effective management of invasive species in coastal wetlands requires accurate knowledge of the spatial distribution of salt marsh vegetation. Remote sensing is one of the most efficient methods for monitoring the physical environment, particularly for highly dynamic and extensive landscapes like coastal wetlands and tidal flats (Phinn et al. 1999, Silvestri et al. 2003). In contrast to a field-based survey, remote sensing imagery can be acquired for all habitats, over a larger spatial area, and in a shorter period of time (Underwood et al. 2003). However, mapping salt marsh vegetation at the species level with traditional remote sensing is still challenging due to its fewer spectral channels and coarse spatial resolution. Vegetation patches in fragmented wetlands are usually smaller than the spatial extent of traditional satellite imagery pixels, and associated bare ground fractions and sediments may vary considerably in space and time, contributing to the mixed pixel problem (Townshend et al. 2000, Okin et al. 2001). Therefore, the spatial scale of remote sensing data suitable for salt marsh vegetation mapping should not exceed a few meters. Airborne hyperspectral imagery with high spatial and spectral resolution offers an enhanced potential for discriminating salt marsh species (Underwood et al. 2003, Aritgas and Yang 2004).

Although the nonunique nature of spectral responses in vegetation makes it unlikely that the separation of vegetation species will be perfect (Cochrane 2000), the small difference of spectral reflectance on different wavelengths of the electromagnetic spectrum is still the best way to discriminate vegetation types (Schmidt and Skidmore 2003). In recent years, efforts have been made in classifying vegetation types using hyperspectral remote sensing (Eastwood et al. 1997, Silvestri et al. 2002, Kokaly et al. 2003). However, few studies have focused on the mapping of salt marsh vegetation and invasive species in fragmented coastal wetlands. Schmidt and Skidmore (2003), for example, examined and tested the differences of reflectance spectra of 27 vegetation types in the Dutch Waddenzee ecosystem and concluded that salt marsh vegetation types may be identified from well-calibrated hyperspectral imagery using a spectral library measured in the field. In another study, Silvestri et al. (2003) used a linear unmixing technique to separate salt marsh vegetation communities in a Venice lagoon in northeastern Italy. Authors also tested spectral discrimination of salt marsh species in the Meadowlands in northern New Jersey in a previous study of mapping vigor gradients of salt marsh vegetation (Artigas and Yang 2005). The objectives of this study were (1) to investigate the use of hyperspectral imagery in mapping salt marsh vegetation in a coastal wetland, and (2) to evaluate the methods of endmembers selection in hyperspectral image classification.

3.2 STUDY AREA AND DATA

The New Jersey Meadowlands is located in northeastern New Jersey, approximately three miles west of New York City. It is a mixture of highly developed residential and industrial land uses interspersed among expanses of landfills, marsh grass fields, tidal wetlands creeks, mudflats, and rivers (Figure 3.1). There are approximately

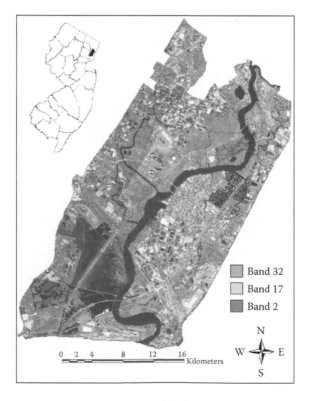

FIGURE 3.1 False color composite (RGB = 32:17:2) of AISA hyperspectral imagery of the New Jersey Meadowlands. **(See color insert after p. 162.)**

34 km² of wetlands and open water within the Meadowlands, and 12 km² of salt marsh vegetation including high marsh species *Patens* (*Spartina patens*) and *Distichlis* (*Distichlis spicata*), and low marsh species *Spartina* (*Spartina alterniflora*). The invasive species, *Phragmites*, outcompetes the native species and results in thick stands of up to 4 ~ 5 meters high on tide-restricted areas, higher elevation dredge spoil islands, and tidal creek banks and levees.

Hyperspectral imagery of the New Jersey Meadowlands was acquired on 11 October 2000 using Airborne Imaging Spectroradiometer for Applications (AISA). AISA is a solid-state, push-broom instrument capable of collecting data within a spectral range of 430 to 900 nm in up to 286 spectral channels. The sensor was configured for 34 spectral bands from 452 to 886 nm and 20 degrees of field of view (FOV) at 2,500 m altitude, corresponding to a swath width of 881.6 m and pixel size of 2.5 × 2.5 m. Atmospheric conditions on the day of image acquisition were clear sky with 660 Watts/m² of solar irradiation at ground level, 55% of relative humidity, and 18°C of surface temperature.

In situ reflectance spectra of dominant salt marsh vegetation were collected from relatively homogeneous 10 × 10 m plots at six different locations using a FieldSpec® Pro Full Range Spectroradiometer from Analytical Spectral Devices (ASD 1997). The spectroradiometer was configured to an eight degree of FOV at a height of 1.5 m, which gives a 0.26-m diameter ground extent. Field spectra measurements were collected under clear skies within 1.5 hours of high sun and referenced to a Spectralon® white reference panel before and after each sampling period. All profile measurements were calculated by averaging 25 samples to reduce the noise.

3.3 METHODS

Both geometric and brightness corrections were conducted on the AISA image. More than 200 ground control points (GCPs) were selected on both AISA imagery and reference orthophoto for geometric correction. Each strip was warped to New Jersey State Plane map projection (NAD83) using a nearest-neighbor resampling algorithm with an average root-mean-square error (RMSE) of ±0.88 pixel (or ±2.2 m). Cross-Track Illumination Correction Function was used to eliminate brightness distortion between strips (Research Systems, Inc. [RSI] 2003). After corrections, 22 AISA strips were mosaicked into a single seamless image and then subset to the Meadowlands district (Figure 3.1). Considering that not all 34 spectral bands contributed useful information, a minimum noise fraction (MNF) rotation was applied to reduce the computation (Underwood et al. 2003). Based on the MNF output graph of eigenvectors, the first 15 MNF bands were chosen for further analysis and vegetation mapping.

Normalized difference vegetation index (NDVI) is a measure of density and vigor of green vegetation growth using the spectral reflectivity of solar radiation and is usually derived from the following equation (Carlson and Ripley 1997):

$$NDVI = \frac{\left(\alpha_{nir} - \alpha_{red}\right)}{\left(\alpha_{nir} + \alpha_{red}\right)} \tag{3.1}$$

where α_{red} and α_{nir} represent surface reflectance averaged over ranges of wavelengths in the red (500 to 700 nm) and near infrared (700 to 900 nm) regions of the electromagnetic spectrum, respectively.

Band 12 around 600 nm and band 28 near 800 nm of the AISA image were selected to calculate the NDVI image in this study. Pixels with NDVI values less than 0.3 were selected as a mask to remove nonvegetated pixels such as impervious surfaces, open water, and mud from the image so that the retained image contains only vegetated pixels. Other vegetated pixels, such as those of upland forest stands, were then removed using a second mask generated from a land use map.

Two methods were used to select endmembers spectra in this study. One was to select endmembers from a spectral library, which consists of 17 spectra of salt marsh surfaces collected in the field. The reflectance spectra of five selected endmembers from the spectral library are shown in Figure 3.2a. The other method was to select endmembers from the AISA image by locating pure pixels areas with monospecific vegetation through direct field inspection (Kokaly et al. 2003). Six endmembers selected in this way include high marsh, *Spartina* pure, *Spartina* mixture, *Spartina/Phragmites* stunted, *Phragmites* big flower, and *Phragmites* small flower (Figure 3.2b).

After endmembers selection, a supervised classifier, the spectral angle mapper (SAM), was used to perform image classification on 15 MNF bands. The SAM algorithm determines the similarity of two spectra by calculating the *spectral angle* between each spectrum in the image and the endmembers in *n* dimensions, treating them as vectors in a space with dimensionality equal to the number of bands (Kruse et al. 1993). The maximum angles between 0.05 and 0.2 were used in the SAM classification, and the angle that resulted in the highest map accuracy was chosen for final vegetation mapping.

3.4 RESULTS AND DISCUSSION

Figure 3.2a shows typical reflectance spectra of salt marsh vegetation species measured in the field. The two high marsh species, *Distichlis* and *Patens*, are difficult to separate in the visible region, but they are separable spectrally in the near-infrared region. In addition, native vegetation species can be easily separated from the invasive *Phragmites*, while the separation between the two native species *Patens* and *Spartina* is not easy. Figure 3.2b shows the typical reflectance spectra of salt marsh vegetation derived from the AISA image by field inspection. *Distichlis* and *Patens* were combined into one category as high marsh, and low marsh was divided into *Spartina* pure, *Spartina* mixture, and *Spartina* stunted. Marsh reed was also grouped into two categories: *Phragmites* big flower (higher than or equal to six feet) and small flower (lower than six feet).

Some differences exist between field-measured and image-derived endmembers spectra. First, except for stunted species, reflectance spectra of all other species derived from the image are lower than those measured in the field. Second, the reflectance of field-measured endmembers increased in the order of *Distichlis*, *Spartina*, and *Phragmites* in the near-infrared region, while the reflectance of image-derived endmembers increased in the order of *Spartina*, *Phragmites*, and *Distichlis* in the

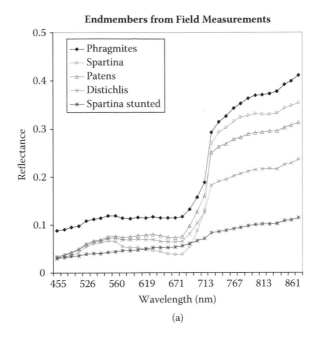

(a)

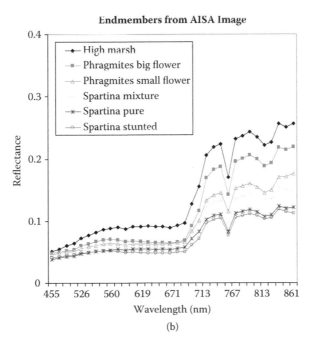

(b)

FIGURE 3.2 Endmembers spectra derived from (a) field measurements with spectro-radiometer, and (b) the area of pixels with monospecific vegetation in AISA imagery through field inspection.

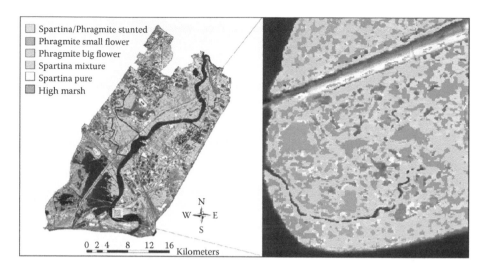

FIGURE 3.3 Map of salt marsh vegetation in the New Jersey Meadowlands with the insert of the Bend, which shows the detailed distribution of salt marsh species in six categories. (**See color insert after p. 162.**)

same region. This is most likely because the locations of training sites selected from the image are different from those measured in the field. Another possible reason is due to the atmosphere that was not completely accounted for through the atmospheric correction. Further atmospheric correction is needed to remove the water vapor absorption effects, which exist in 760 nm and 830 nm in Figure 3.2b.

The spatial distribution of salt marsh vegetation in the Meadowlands is presented in Figure 3.3. *Phragmites* occupies approximately 80% of the entire salt marsh vegetation in the Meadowlands, *Spartina* and its mixture 10%, high marsh only 1%, and the remaining 10% are stunted *Spartina/Phragmites*. The accuracy assessment was conducted first on three main salt marsh surfaces: high marsh (*Distichlis* and *Patens*), low marsh (*Spartina* and its mixtures), and marsh reed (*Phragmites* big and small flower) using 38 ground truth points in the southern Meadowlands. The results showed that the method using image-derived endmembers performed better, 85% overall accuracy (kappa = 0.76) compared to 75% (kappa = 0.53) for the method using field-collected endmembers. In accuracy assessment conducted for five vegetation species, image-derived endmembers also resulted in higher accuracy than field-collected endmembers. The method using image-derived endmembers resulted in 63.2% overall accuracy (kappa = 0.53) for mapping five species of salt marsh vegetation and 71.4% producer's accuracy for mapping invasive *Phragmites* (Table 3.1).

Close examination of the salt marsh vegetation map revealed some possible errors in salt marsh vegetation mapping. We found that many places are a mixture of *Phragmites* and native species, which make the spectral discrimination difficult. *Phragmites*-stunted and *Spartina*-stunted surrounding mud are also difficult to distinguish due to the high moisture content. Ground truth points collected in the southern portion of the Meadowlands may also generate bias for the accuracy in the northern part of the Meadowlands. More ground truth points are needed to perform a more reliable assessment for the entire Meadowlands.

TABLE 3.1

Error matrix of the classified vegetation map derived from the AISA image using image-derived endmembers in the New Jersey Meadowlands.

	Reference Data					
Classification	HM marsh	SP pure	SM mixture	BF Phragmites	SF Phragmites	Row Total
High marsh	11	0	0	0	0	11
Spartina pure	0	1	0	0	0	1
Spartina mix	2	1	2	2	1	8
Big flower *Phragmites*	3	0	1	5	1	10
Small flower *Phragmites*	0	0	0	0	5	5
Column total	16	5	3	7	7	38

Overall accuracy = 63.16 %   Kappa coefficient = 0.5309

Producer's Accuracy (%)		User's Accuracy (%)	
High marsh (HM)	68.8	High marsh (HM)	100
Spartina pure (SP)	20.0	*Spartina pure (SP)*	100
Spartina mix (SM)	66.7	*Spartina mix (SM)*	25
Big flower (BF)	71.4	Big flower (BF)	50
Small flower (SF)	71.4	Small flower (SF)	100

3.5 CONCLUSION

This study describes a method for mapping salt marsh vegetation and invasive species using hyperspectral AISA imagery. Generally, the method using image-derived endmembers resulted in higher mapping accuracy than the method using field-collected endmembers. More attention needs to be given to the atmospheric effects, which make the spectra derived from the AISA image different from those measured in the field. The results show that by carefully collecting endmembers from the image through field inspection, the SAM method is able to classify the hyperspectral imagery with respect to salt marsh vegetation mapping at the species level with acceptable accuracy. This study will contribute to the knowledge base of land managers by providing improved information concerning spatial distribution and density of salt marsh vegetation in coastal wetlands, which will lead to better understanding and management of invasive species and its native biodiversity.

REFERENCES

Artigas, F. J., and J. Yang. 2004. Hyperspectral remote sensing of habitat heterogeneity between tide-restricted and tide-open areas in New Jersey Meadowlands. *Urban Habitat* 2(1):1–18.

Artigas, F. J., and J. Yang. 2005. Hyperspectral remote sensing of marsh species and plant vigor gradient in the New Jersey Meadowland. *International Journal of Remote Sensing* 26:5209–5220.

ASD (Analytical Spectral Devices). 1997. *Technical Guide*. Boulder, CO: ASD.

Carlson, T. N., and D. N. Ripley. 1997. On the relation between NDVI, fractional vegetation cover, and leaf area index. *Remote Sensing of Environment* 62:241–252.

Cochrane, M. A. 2000. Using vegetation reflectance variability for species level classification of hyperspectral data. *International Journal of Remote Sensing* 21(8):2075–2087.

Eastwood, J. A., M. G. Yates, A. G. Thomson, and R. M. Fuller. 1997. The reliability of vegetation indices for monitoring salt marsh vegetation cover. *International Journal of Remote Sensing* 18(18):3901–3907.

Herrera-Silveira, J. A., F. A. Comin, N. Aranda-Cirero, L. Troccoli, and L. Capurro. 2004. Coastal water quality assessment in the Yucatan Peninsula: Management implications. *Ocean & Coastal Management* 47(11–12):625–639.

Ko, J. Y., and J. D. Day. 2004. A review of ecological impacts of oil and gas development on coastal ecosystems in the Mississippi Delta. *Ocean & Coastal Management* 47(11–12):597–623.

Kokaly, R. F., D. G. Despain, R. N. Clark, and K. E. Livo. 2003. Mapping vegetation in Yellowstone National Park using spectral feature analysis of AVIRIS data. *Remote Sensing of Environment* 84:437–456.

Kruse, F. A., A. B. Lefkoff, and J. B. Dietz. 1993. Expert system–based mineral mapping in northern Death Valley, California/Nevada using the Airborne Visible/Infrared Imaging Spectrometer (AVIRIS). Special issue on AVIRIS, *Remote Sensing of Environment* 44:309–336.

Okin, G. S., D. A. Roberts, B. Murray, and W. J. Okin. 2001. Practical limits on hyperspectral vegetation discrimination in arid and semiarid environments. *Remote Sensing of Environment* 77:212 –225.

Phinn, S. R., D. A. Stow, and D. Van Mouwerik. 1999. Remotely sensed estimates of vegetation structural characteristics in restored wetlands, Southern California. *Photogramm. Eng. Remote Sensing* 65(4):485–493.

RSI (Research Systems, Inc.). 2003. ENVI 4.0 Users Guide.

Schmidt, K. S., and A. K. Skidmore. 2003. Spectral discrimination of vegetation types in a coastal wetland. *Remote Sensing of Environment* 85:92–108.

Silvestri, S., M. Marani, and A. Marani. 2003. Hyperspectral remote sensing of salt marsh vegetation, morphology and soil topography. *Physics and Chemistry of the Earth* 28: 15–25.

Silvestri, S., M. Marani, J. Settle, F. Benvenuto, and A. Marani. 2002. Salt marsh vegetation radiometry: Data analysis and scaling. *Remote Sensing of Environment* 2:473–482.

Townshend, J. R. G., C. Huang, S. N. V. Kalluri, R. S. Defries, S. Liang, and K. Yang. 2000. Beware of per-pixel characterization of land cover. *International Journal of Remote Sensing* 21(4):839–843.

Underwood, E., S. Ustin, and D. DiPietro. 2003. Mapping nonnative plants using hyperspectral imagery. *Remote Sensing of Environment* 86:150–161.

4 Carex Mapping in the Poyang Lake Wetland Based on Spectral Library and Spectral Angle Mapping Technology

Shuisen Chen, Liangfu Chen, Xiaobo Su,
Qinhuo Liu, and Jian Li

4.1 INTRODUCTION

Located in the northern Jiangxi Province, the Poyang Lake wetlands have played an important role in controlling floods, providing habitation, purifying toxicants, and adjusting climate. This valuable resource has been severely depleted in recent years due to excessive exploitation, resulting in the decline of the wetland's functions and self-restoration ability (Zhang 2004, Zhu et al. 2004). The inaccessibility of the area makes it difficult and expensive to monitor and assess the dynamics of this freshwater lacustrine system. Thus, remote sensing technology has become a necessary and efficient tool for this task (Nepstad et al. 1999). Remotely sensed data have been extensively used to map land cover (including wetland areas) for the purpose of environmental conservation, for example, identifying areas demanding protection and monitoring important habitats (Steininger et al. 2001, Turner et al. 2003).

Remotely sensed data has previously been used for acquiring information about the environment and resources of the Poyang Lake wetland, including specific characteristics analysis of healthy vegetation (Xiaonong et al. 2002) and multitemporal analysis of land cover changes. However, the wetland mapping in Poyang Lake was not completely automatic (Jian et al 2001, Zhao et al. 2003), which hindered quick acquisition of the biological resource information necessary for utilization and protection of the wetland. The purpose of this study is: (1) to develop an automatic method for mapping the wetland using Landsat ETM+ or TM images, and (2) to map the distribution of *Carex* based on the spectral library and the spectral angle mapping approach, and (3) to assess the accuracy of the presented approach.

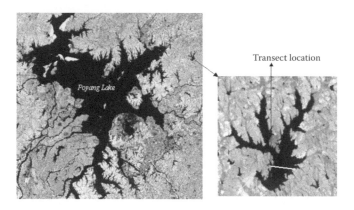

FIGURE 4.1　Location of reference reservoir transect in the lake area.

4.2 DATA AND METHODS

4.2.1 Data

Two Landsat images, TM (July 15, 1989, flood season), and ETM+ (December 10, 1999, dry season), were used in this study. The first was used for automatic mapping of the flooding area and the second for the extraction of beach vegetation information within the Poyang Lake cofferdam. Before subsetting images, radiometric calibration was performed to transform the digital number (DN) values into image reflectance.

4.2.2 Flooding Area Extraction

The flood season image was used to create a mask of two primary classes—water and land. The flood plain extent of Poyang Lake during the flood period was used for the automatic mapping of the lake beach vegetation distribution area in the low water season, limiting the *Carex* mapping area within the lake beach wetland and water area. The identified flood plain area from the July image was used as a mask to obtain the lake beach wetland and water area from the December image, that is, our study area. From Figures 4.1 and 4.3, it is obvious that the water body and land have different reflectance values. This allows the setting of a threshold to distinguish the water body from the land using reflectance values. As a comparison, in previous research (Niu and Zhao 2004), only the DN values of Landsat imagery were used, which reduced the precision of wetland vegetation mapping using spectral information. Based on the statistical analysis of reflectance of the water body and land region on Band 4 (Figures 4.2 and 4.4), a threshold value of 0.12 was applied to distinguish water and nonwater areas on the image of July 15, 1989, because Band 4 (near-infrared band) provided the best differentiation between water and land. Image pixels with a reflectance value less than 0.12 were classified as a water body (including both lake and reservoir waters), and the rest as land. It can be seen from Figure 4.4 that the reflectance of lake water is smaller than 0.12.

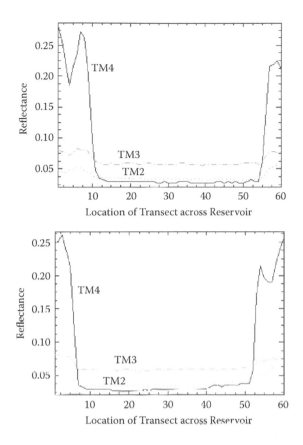

FIGURE 4.2 Reflectance transect of ETM⁺ Band 4 across a reservoir.

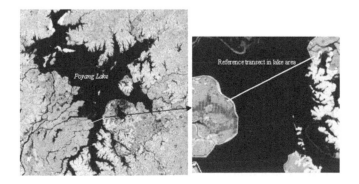

FIGURE 4.3 Location of the lake water reference transect in the lake area.

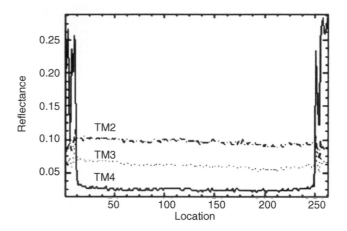

FIGURE 4.4   Reflectance transect of ETM⁺ band 4 across Poyang Lake.

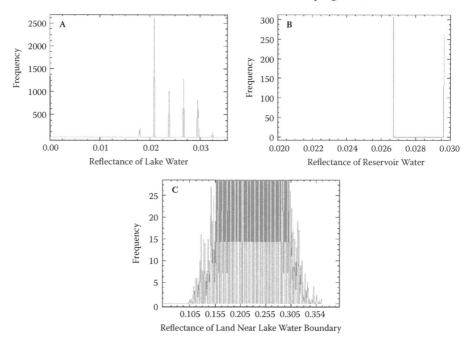

FIGURE 4.5   Spectral angles of lake water endmembers, lake water, reservoir, and land.

4.2.3 Spectral Angle Mapping (SAM)

One of the most commonly used methods in spectrometry is the comparison of spectral angles among different land covers. Figure 4.5 illustrates the difference in spectral angles among natural materials (lake water, reservoir water, land) spectra and the lake water endmembers, with the analysis results in Table 4.1. Traditionally, the spectral distance and probability-based classification methods do not consider the linear scaling of overall reflectance patterns. Spectral angle mapping techniques, on

TABLE 4.1

Various ranges of different water and nonwater bodies in reflection.

Sample land cover	Minimum	Maximum	Scope of most pixels
Lake water	0.015	0.0355	100%
Reservoir water	0.0267	0.296	100%
Nonwater body	0.1045	0.3650	0.1403–0.3650 (98.7%)
			0.1245–0.3650 (99.6%)

the other hand, incorporate linearly scaled reflectance patterns to avoid the misclassification of land use and land covers, which are the linearly scaled versions of a particular reflectance pattern.

The angle that defines a spectral signature or class does not change, and the vectors forming the angle from the origin delineate and contain all possible positions for the spectra (Sohn and Rebello 2002). These parameters encompass all the possible combinations of illumination for the spectra. Changes in reflectance due to the effect of illumination are still within the class angle (Sohn et al. 1999). The fact that the spectra of the same type are approximately linearly scaled versions of one another due to illumination and topographic variations is utilized to achieve accurate classification results (Sohn et al. 1999). Because the spectral angle classifier utilizes the shape of the pattern for the clustering and classification of multispectral image data (Sohn and Rebello 2002, Luo and Chen 2002, An et al. 2005), the analyst's ability to relate field information to spectral characteristics and spectral shape patterns for different land cover and land use types is an important factor for achieving accurate mapping results.

In this study, spectral angle–based statistical analysis could better quantify *Carex* vegetation than other wetland types. It is more convenient and useful than traditional supervised and unsupervised methods of classification.

4.3 RESULTS AND DISCUSSION

The comparison of reflectance range and frequency between different land covers helps to distinguish various water bodies from land in the study area (Figure 4.5). Table 4.1 further explains the reflective difference between water bodies and land. The multispectral composite image of water bodies in the study area indicates that the extent of light-colored water represents more turbid lake water (Figure 4.6); Figure 4.7 depicts the spatial distribution of the reflection for different water bodies. Figure 4.8 is a vector map of water body distribution within the boundary of 14 lakeshore towns in Poyang Lake wetland area. Figures 4.8 and 4.9 show the distribution of dense *Carex* based on the spectral angles of image pixels with reference to the *Carex* endmember. The total area of *Carex* is 166 km². According to the formula for calculating landscape fragmentation index:

$$C = \Sigma Ni/A$$

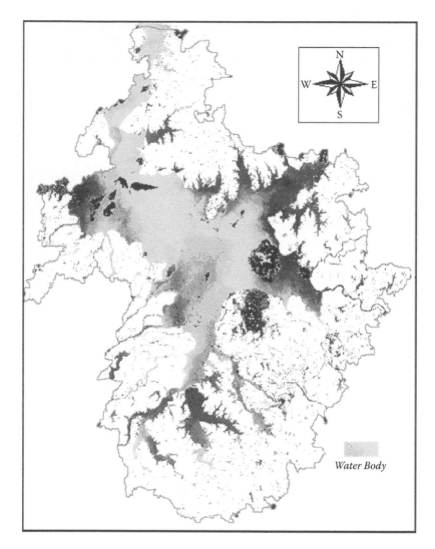

FIGURE 4.6 Multispectral component image of extracted water body in study area (R, G, B = TM 4, 3, 2). (**See color insert after p. 162.**)

where *C* is the landscape fragmentation index of *Carex,* *ΣNi* stands for the total number of *Carex* landscape type polygons, and *A* is the total area of the *Carex* landscape.

The fragmentation index of the *Carex* landscape in the Poyang Lake wetland area is 0.6041 (0 represents landscape that has not been depleted, and 1 represents landscape that has been completely destroyed), which is higher than the Honghu Lake wetland, the other important wetland in China, with a landscape fragmentation index of 0.4207 (Wang et al. 2005). This figure indicates that the landscape in Poyang Lake is influenced by human activities more severely than that of the Honghu Lake wetland. The area of Poyang Lake's water body within the boundary of 14 lakeshore towns

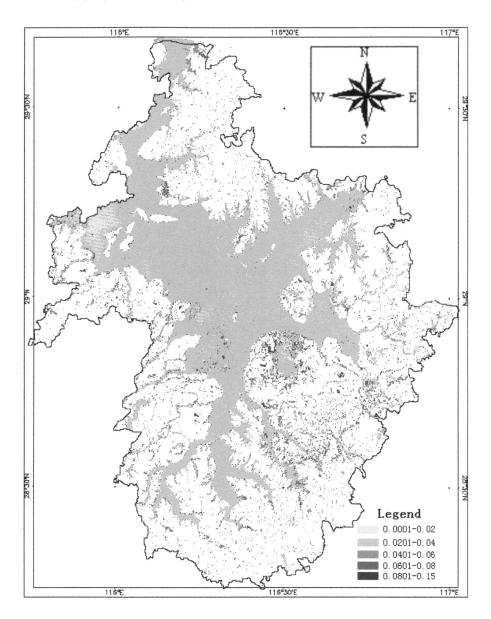

FIGURE 4.7 Water bodies of different reflection in the study area. (**See color insert after p. 162.**)

in Figures 4.6 through 4.8 is 4,209 km^2 while the central lake water area is 3,340 km^2 (Figure 4.10). The precision of the dense *Carex* mapping was validated by the field of investigation (Figure 4.11). According to the 11 field sites, there were 10 sites that had growth in dense *Carex,* with the remaining site having growth of *Arte misia selengensis.* However, there was *Carex* growing under the *Arte misia selengensis.* A high precision of 91% for *Carex* mapping was achieved.

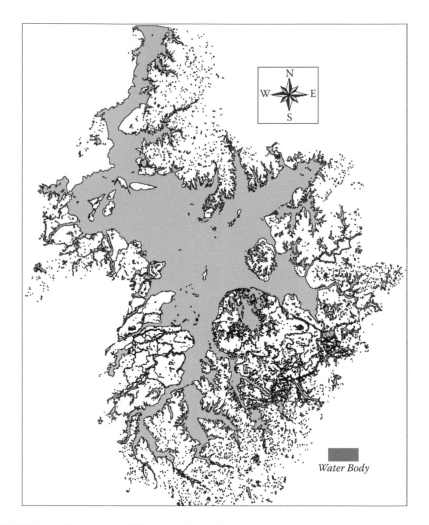

FIGURE 4.8   Vector map of Poyang Lake wetland.

4.4 CONCLUSION

The automatic and quick mapping method for discriminating land and water, and different turbidities of water, has proved to be an effective tool for wetland monitoring. Specifically, it was possible to accurately detect the dense *Carex* by the spectral angle mapping approach based on the comparison of the reflection of *Carex* endmembers and the image pixels. The method used to extract the flood plain could also be used to automap the other wetland types. This study shows that multitemporal images with their near-infrared reflection of water bodies and spectral library, are valid choices for automatic and quick wetland classification.

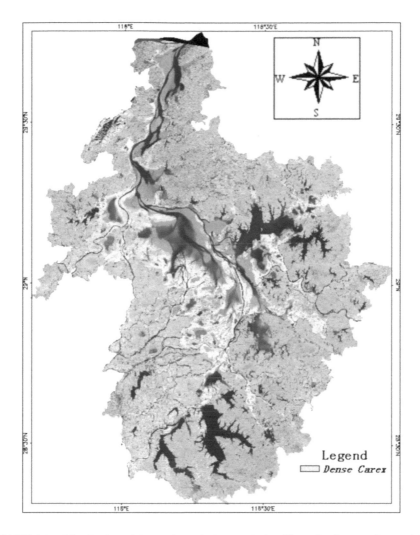

FIGURE 4.9 Distribution of dense *Carex* in Poyang Lake. **(See color insert after p. 162.)**

ACKNOWLEDGMENTS

This research was supported by the open fund of the Key Lab of Poyang Lake Ecological Environment and Resource Development of the Chinese Ministry of Education housed in Jiangxi Normal University, China's 863 High-Tech Research Plan Project (No: 2002AA130010), and the 2005 Science and Technology Plan Fund of Jiangxi Province. Thanks also to two anonymous reviewers for their helpful comments and suggestions.

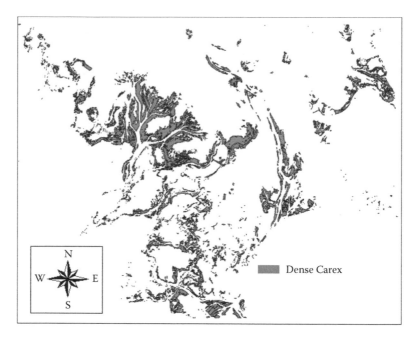

FIGURE 4.10   Vector map of *Carex* distribution focused on the central Poyang Lake wetland area.

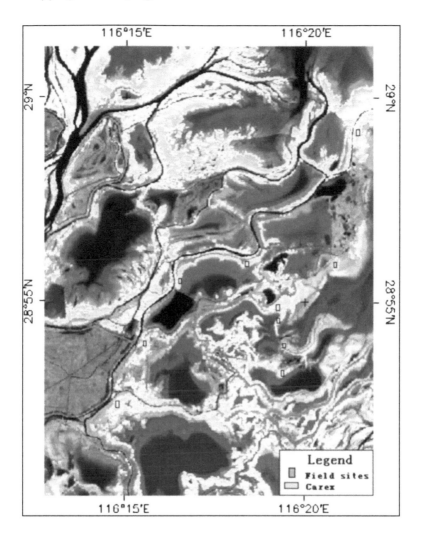

FIGURE 4.11 Sites of field investigation in the study area. (**See color insert after p. 162.**)

REFERENCES

An, Bin, Shu-hai Chen, and Wei-dong Yan. 2005. Application of SAM algorithm in multi-spectral image classification [in Chinese]. *Chinese Journal of Stereology and Image Analysis* 11(11):55–61.

Jian, Yongxing, Rendong Li, Jianbo Wang, and Jiakuan Chen. 2001. *Acta Phytoecologica, Sinica* 25(5):581–587.

Lan, Tianwei. 2004. Poyang Lake wetland is formally applied to be protected as world's natural heritage. http://news.xinhuanet.com/house/2004-11/04/content_2175744.htm.

Luo, Yu-xia, and Huan-wei Chen. 2002. Comparing degree classification with distance classification—with salt soil as an example [in Chinese]. *Remote Sensing for Land and Resources* 2:46–48.

Nepstad, D. C., A. Verissimo, A. Alencar, C. Nobre, E. Lima, P. Lefebvre, P. Schlesinger, C. Potter, P. Moutinho, E. Mendoza, M. Cochrane, and V. Brooks. 1999. Large-scale impoverishment of Amazonian forests by logging and fire. *Nature* 398:505–508.

Niu, Ming-xiang, and Geng-xing Zhao. 2004. Study on remote sensing techniques on wetland information extracting in Nansihu area. *Territory and Natural Resources Study* 4:51–53.

Sohn, Y., E. Moran, and F. Gurri. 1999. Deforestation in north-central Yucatan (1985–1995): Mapping secondary succession of forest and agricultural land use in Sotuta using the cosine of the angle concept. *Photogrammetric Engineering and Remote Sensing* 65(8):947–958.

Steininger, M. K., C. J. Tucker, J. R. G. Townshend, T. J. Killeen, A. Desch, V. Bell, and P. Ersts. 2001. Tropical deforestation in the Bolivian Amazon. *Environmental Conservation* 28:127–134.

Turner, W., S. Spector, N. Gardiner, M. Fladeland, E. Sterling, and M. K. Steininger. 2003. Remote sensing and biodiversity science and conservation. *Trends in Ecology and Evolution* 18:306–314.

Wang, Xi, Xianyou Ren, and Fei Xiao. 2005. Remote sensing and GIS-based landscape structure analysis of Honghu Lake wetland, www.sdinfo.net.cn/xinxizhuanti/2005/xxzt-78.html.

Wessels, K. J., R. S. De Fries, J. Dempewolf, L. O. Anderson, A. J. Hansen, S. L. Powell, and E. F. Moran. 2004. Mapping regional land cover with MODIS data for biological conservation: Examples from the Greater Yellowstone Ecosystem, USA and Pará State, Brazil. *Remote Sensing of Environment* 92(1):67–83.

Xiaonong, Zhou, Lin Dandan, Yang Huiming, Chen Honggen, Sun Leping, Yang Guojing, Hong Qingbiao, Leslie Brown, and J. B. Malone. 2002. Use of Landsat TM satellite surveillance data to measure the impact of the 1998 flood on snail intermediate host dispersal in the lower Yangtze River Basin. *Acta Tropica* 82(2):199–205.

Youngsinn, Sohn, and N. Sanjay Rebello. 2002. Supervised and unsupervised spectral angle classifiers. *Photogrammetric Engineering and Remote Sensing* 68(12):1271–1280.

Zhao, Xiaomin, Mengxian Yuan, and Hu Wang. 2003. Study on remote sensing investigation and comprehensive utilization of low-grassland in Poyang Lake Region. *Acta Agriculturae Universitatis Jiangxiensis* 25(1):84–87.

Zhang, Juntao. 2004. Elementary accounting of resources and environment loss in Poyang Lake wetland [in Chinese]. *Statistical Research* 8:9–12.

Zhu, Lin, Yingwei Zhao, and Liming Liu. 2004. Protective utilization and function estimate of wetlands ecosystem in Poyang Lake [in Chinese]. *Journal of Soil and Water Conservation* 18(2):196–170.

Part II

Wetland Hydrology and Water Budget

5 Change in the Southern U.S. Water Demand and Supply over the Next Forty Years

Steven G. McNulty, Ge Sun, Erika C. Cohen, and Jennifer A. Moore Myers

5.1 INTRODUCTION

Water shortages are often considered a problem in the western United States, where water supply is limited compared to the eastern half of the country. However, periodic water shortages are also common in the southeastern United States due to high water demand and periodic drought. Southeastern U.S. municipalities spend billions of dollars to develop water storage capacity as a buffer against periodic drought. Buffers against water shortage include the development of water reservoirs and well excavation to mine ancient aquifers. It is important to have good estimates of future water supply and demand to prevent wasting money by creating more reservoir capacity than is needed by a community. Conversely, a lack of water reserve capacity can lead to the need for water restrictions.

Many factors impact the amount of water that is available and the amount of water that is required by a community. Precipitation is the major determinant of water availability over the long term. In addition to precipitation, air temperature and land cover also impact water availability by modifying how much precipitation is evaporated and transpired back into the atmosphere. Finally, ancient aquifers provide a significant proportion of needed water in some parts of the southern United States such as Texas and Arkansas. The water recharge rates for deep aquifers are very low (i.e., hundred of years), so water is essentially mined from these areas. There are also several factors that control water demand. In addition to residential water use, a great deal of water is also required by industry, irrigated farming, and the energy production sectors. In total, domestic and commercial sectors account for only 5% of the surface water use and 10% of the groundwater use. The other 95% and 90% of the surface and groundwater are used by other sectors of the economy. Thermoelectric power generation uses 50% of the surface water, and crop irrigation and livestock use 67% of the groundwater. Not all of the water that is used by each sector is lost to the atmosphere; much is quickly returned to the environment to be used again and again. The proportion of returned water varies from 98% from the thermoelectric sector to a 39% return rate from the irrigation and livestock sectors.

Communities need to accurately assess future water supply and demand if water limitations are to be minimized. However, estimates of average annual water supply and demand will be of limited use in water planning. Water shortages generally only occur during extreme event years such as when water supply may be limited by drought or when water demand is high due to drought and high air temperature. Therefore, this paper examines the sensitivity of the individual factors that influence regional water supply and demand across a range of environmental variability. Some factors such as population are relatively stable from one year to the next, while other variables such as climate can be markedly different among years. Groundwater withdrawal may be relatively stable until the supply is depleted. We will determine how much each variable is likely to influence annual water supply stress and the extent to which water supply stress may vary between 1990 and 2045.

5.2 METHODS

Water supply and demand are the two components required to assess regional water stress. We will first define the variables needed to calculate each water supply stress component and then examine how the variables change from year to year.

5.2.1 CALCULATIONS FOR ESTIMATING WATER SUPPLY

We define water availability as the total potential water supply available for withdrawal for each eight-digit hydrologic unit code (HUC) watershed, expressed in the following formula (equation 5.1):

$$WS = P - ET + GS + \Sigma RF_i \qquad (5.1)$$

where WS = water supply in millions of gallons for each watershed per year; P = precipitation for each watershed in cm per year; ET= watershed evapotranspiration for each watershed in cm per year, calculated by an empirical formula as a function of potential evapotranspiration, precipitation, and land cover types; and GS = historic groundwater use in millions of gallons for each watershed per year. Detailed methods are found in Sun et al. (2005).

Most of the water removed for human use is returned to the environment as return flow (RF). We used RF estimates from each of seven water use sectors (WU_i) including domestic, commercial, irrigation, thermoelectric, industrial mining, and livestock sectors, which were reported by the USGS (United States Geological Survey 1994). The RF was calculated as the historical return flow rate (RFR), expressed as a fraction of the amount of removed water that was returned to the ecosystem, multiplied by the total water use (WU) for each sector, expressed in millions of gallons per year for each watershed.

5.2.2 GROUNDWATER SUPPLY DATA

The USGS has published estimates of national water use since 1950 (Solley et al. 1998). The groundwater term (GS) is the 1990 estimate of groundwater use in millions of gallons of water per year for each eight-digit hydrologic unit code (HUC) watershed. Groundwater is defined as the saturated zone below the subsurface (Solley et al. 1993). The 1990 groundwater supply estimate was incorporated into the 1990 baseline year, 2020 wet year, and 2024 dry year water supply stress scenarios. The groundwater term was dropped for the 2043 dry year and 2045 wet year scenarios to examine the implications of a lack of groundwater availability on water supply stress.

5.2.3 HISTORIC AND PROJECTED CLIMATE DATA

The U.K. Hadley Climate Research Center HadCM2Sul climate change scenario was used to project climate between 1990 and 2045. The southern United States is expected to become generally warmer and wetter under the HadCM2Sul scenario. We compared the impact of a future hot and dry year (Figures 5.1a and 5.1b) and a

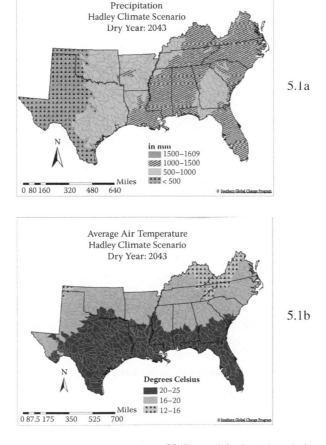

FIGURE 5.1 Predicted hot and dry year (e.g., 2043) precipitation (a) and air temperature (b) distribution, and predicted cool and wet year (e.g., 2045) precipitation (c) and air temperature (d) distribution.

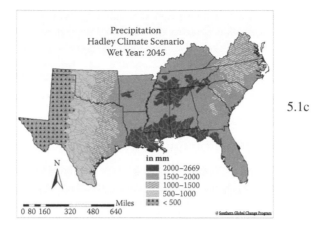

5.1c

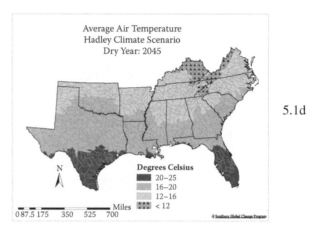

5.1d

FIGURE 5.1 (Continued)

future cool and wet year (Figures 5.1c and 5.1d) with average (circa 1990) historic climate to assess how changes in air temperature and precipitation will impact water supply stress. The climate data were gridded at 0.5° by 0.5° (about 50 km × 75 km) across the continental U.S. (Kittel et al. 1997). Next, 1990 climate data were subset and scaled to the 666 corresponding 8-digit HUC watersheds covering Virginia to Texas (USGS 1994). The same process was repeated for other years used in the water supply stress scenarios. Mean air temperature for 1990 was 24.29°C, while HadC-M2Sul predicts 24.36°C for 2024 and 22.69°C for 2045. Mean precipitation for 1990 was 96.45 cm; HadCM2Sul predicts mean annual values of 78.07 cm for 2024 and 116.68 cm for 2045.

5.2.4 HISTORIC LAND COVER AND LAND USE DATA

Land cover data were used to drive the water yield model for all seven water use sectors and each of the 666 8-digit HUC watersheds modeled in this study. The

1992 Multi-Resolution Land Characterization (MRLC) land cover/land use dataset (http://www.mrlc.gov) was used to calculate the percentage of each vegetation type within each watershed. Land cover was aggregated into five classes including ever-green forest, deciduous forest, crops, urban areas, and water. In this analysis, land use was held constant throughout the assessment period.

5.2.5 Historic and Projected Population Data

Approximately 100 million people live in the 13 southern states (U.S. Census 2002). Population projections at the census block level are available to 2050 (NPA Data Services 1999). We aggregated predicted census block–level population data to the watershed level for each year between 2000 and 2045. Between 1990 and 2045, the population of the 13 southern states was predicted to increase by 94% (NPA Data Services 1999). No new areas of growth within the region were projected, but current urban centers are expected to expand. Rural areas are generally expected to become more densely populated. However, population growth between 1990 and 2045 will not be uniform (Figure 5.2a); percentage change in population between 1990 and 2045 varied from −21% to +602% across the region (Figure 5.2b).

5.2.6 Calculations for Estimating Water Demand

Water demand is as important as water supply in determining if a community is likely to experience recurrent water shortages. Water demand (WD) represents the sum of all water use (WU) by each of the seven water use sectors within a watershed (equation 5.2):

$$WD = \Sigma WU_i \quad i = 1 - 7 \tag{5.2}$$

Annual domestic water use (DWU) for each watershed was predicted by correlat-ing USGS historical water use (in millions of gallons per day per watershed) for the domestic water use sector with watershed population (P, in thousand persons) for the years 1990 and 1995 (equation 5.3):

$$DWU = 0.008706 * P + 1.34597 \quad R^2 = 0.51, n = 666 \tag{5.3}$$

Similarly, irrigation water use (IWU) for each watershed was derived by correlating USGS historical irrigation water use (in millions of gallons per day per watershed) with the irrigation area (IA, in thousands of acres) for the years 1990 and 1995 (equation 5.4):

$$IWU = 1.3714 * IA + 2.06969 \quad R^2 = 0.67, n = 666 \tag{5.4}$$

Currently, we do not have water use models for the other five sectors (i.e., commer-cial, industrial, livestock, mining, and thermoelectric), therefore historic water use data were utilized for future periods.

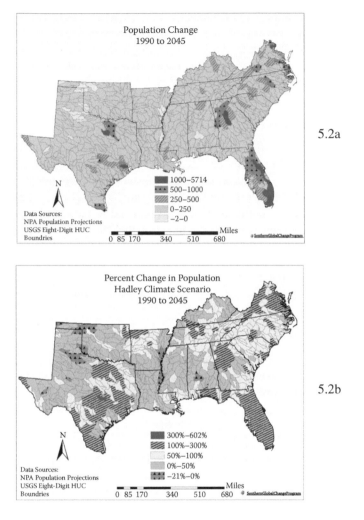

FIGURE 5.2 The spatial and temporal change in southern U.S. population between 1990 and 2045 expressed as absolute change (a) and percentage change (b).

5.2.7 Calculations for Estimating the Water Supply Stress Index (WASSI)

We have defined a water supply stress index (WASSI) by dividing water supply by water demand (equation 5.5). Comparing the WASSI between two points in time results in a WASSI ratio (WASSIR, equation 5.6). The WASSI was used to quantitatively assess the relative magnitude in water supply and demand at the 8-digit HUC watershed scale. The WASSIR was used to assess the relative change in the WASSI between the baseline scenario (SI_1) and one of the other scenarios (SI_x) described later. Positive WASSIR values indicate increased water stress and negatives values indicate reduced water stress compared to historical conditions (scenario 1):

$$WASSI_x = \frac{WD_x}{WS_x} \qquad (5.5)$$

and

$$WASSIR_x = \frac{WASSI_x - WSS_1}{WASSI_1} \qquad (5.6)$$

where x represents one of six simulation scenarios described below.

5.2.8 WATER SUPPLY STRESS SCENARIOS

Six scenarios were developed, each examining the impacts of changing population, climate, and ground water supply on annual WASSI values. Changes in the water supply portion of the WASSI term were addressed in three ways. First, we chose two dry years (2024 and 2043) and two wet years (2020 and 2025) to compare to the historic climate base year of 1990. Second, ground water is a finite resource and given the current rate of usage, it is possible that ground water may be completely depleted in some areas during the next 40 years. We tested the impact of the loss of ground water supply on water supply stress by examining wet and dry years with and without ground water inputs into the WASSI model. Finally, population impacts on water supply stress were examined by comparing wet and dry years in the 2020s and 2040s with the baseline water supply stress year (i.e., 1990).

5.2.8.1 Scenario 1: Small Population Increase—Wet Year (2020)

This scenario used 2020 population projections that predicts above average precipitation for the region compared to 1990 values. Groundwater withdrawal was held constant at 1990 levels.

5.2.8.2 Scenario 2: Small Population Increase—Dry Year (2024)

This scenario used 2024 population projections. 2024 is predicted to have below average precipitation compared to 1990 levels, thus decreasing water supply. Groundwater withdrawal was held constant at 1990 levels.

5.2.8.3 Scenario 3: Large Population Increase—Wet Year (2045)

This scenario used 2045 population projections, and it is predicted to have above average precipitation compared to 1990 levels. Groundwater withdrawal was held constant at 1990 levels.

5.2.8.4 Scenario 4: Large Population Increase—Dry Year (2043)

This scenario used 2043 population projections, that is predicted to have below average precipitation compared to 1990 levels. Groundwater withdrawal was held constant at 1990 levels.

5.2.8.5 Scenario 5: Large Population Increase—Wet Year (2045), No Groundwater Supply (GS)

This scenario used 2045 population projections that is predicted to have above average precipitation compared to 1990 levels. However, groundwater supplies may be exhausted in some areas by 2045. Therefore, we removed groundwater for the entire region as a water supply source to the WASSI model in this scenario.

5.2.8.6 Scenario 6: Large Population Increase—Dry Year (2043), No Groundwater Supply (GS)

This scenario used 2043 population projections that is predicted to have below average precipitation compared to 1990 levels. In addition to the increased population and below average year precipitation, the groundwater supply was removed from the WASSI model across the region in this scenario.

5.3 RESULTS AND DISCUSSION

The results section is divided into three parts: (1) climate controls on water supply stress, (2) population and other water use sector controls on water supply stress, and (3) ground water supply controls on water supply stress.

5.3.1 CLIMATE CONTROLS ON THE WASSI

Historically, annual precipitation and air temperature vary widely across the region: central Texas averages less than 70 cm of precipitation per year, while parts of the Gulf coast and southern Appalachians receive almost 200 cm. Average annual air temperature is roughly inversely proportional to latitude within the region. Annual precipitation and air temperature are the most important determinants of water loss by evapotranspiration and thus water yield across the southern United States (Lu et al. 2003). Therefore, the Appalachians and the Gulf coast had the highest water availability, while the lowest was found in semiarid western Texas.

Irrigation and thermoelectric sectors were the two largest water users, followed by domestic livestock and industrial users. Consequently, the western Texas region, which had a lot of irrigated farmland and limited water supplies, had the highest WASSI for both wet and dry years (Figures 5.3a–d). Other areas with WASSI values indicating stress included southern Florida, southern Georgia, and Mississippi valley areas with a high percentage of irrigated land (relative to the total land area). Several isolated watersheds in high-precipitation regions east of the Mississippi River also showed high water stress, primarily due to thermoelectric water use.

Compared to the baseline conditions of 1990, the WASSI in 2020 (Figure 5.3a) and 2045 (not shown) were projected to decrease due to a moderate increase in air temperature and a large increase in precipitation. The WASSI in 2024 (Figure 5.3b) was projected to increase due to a moderate temperature increase and moderate decrease in precipitation compared to 1990 levels. As a result, the average regional WASSIR value (compared to 1990) decreased by 5% during the wet year of 2020 (Figure 5.3c), but increased by 22% for the dry year of 2024 (Figure 5.3d) and 66%

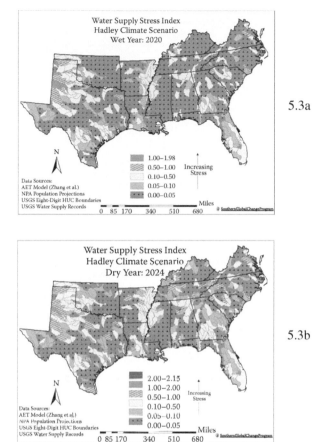

FIGURE 5.3 Climate change impacts on the southeastern U.S. water supply stress (WASSI) during the wet year of 2020 (a) and the dry year of 2024 (b). Change in the water supply stress ratio (WASSIR) between the 1990 baseline WASSI and the wet year WASSI of 2020 (c) and the dry year WASSI of 2024 (d).

for the dry year of 2043 (not shown). The WASSIR again decreased for the dry year of 2045 by 4% (not shown).

5.3.2 POPULATION AND OTHER WATER USE SECTOR CONTROLS ON THE WASSI

Water demand by the domestic water use sector is directly related to population, as demonstrated in Equation 5.3. Population centers that are projected to expand dramatically over the next 40 years (e.g., Atlanta, Georgia; Dallas, Texas; Raleigh-Durham, North Carolina; and northern Virginia) will see up to 200% increases in domestic water use. Therefore, population growth may be responsible for increasing the WASSI by more than 70% in watersheds containing relatively large increases in population, but population change will have little impact (<5%) on the regional-scale WASSI.

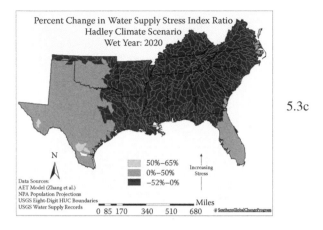

5.3c

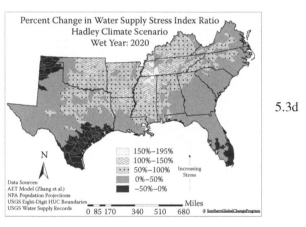

5.3d

FIGURE 5.3 (Continued)

The mean regional WASSI increased from 0.11 (Figure 5.4a) to 0.12 (Figure 5.4b), despite a 30% increase in southeastern U.S. population between 2024 and 2043. Population changes had little impact on the WASSI when compared to that of the interannual variation in climate and potential loss of groundwater reserves. A doubling of local populations around metropolitan areas will have a limited impact on interannual WASSI variability. Even in heavily populated areas, residential and commercial water use represent small segments of total water demand, but cities do affect water quality. As the population increases, costs for water treatment and acquisition increase; water conservation may be important for reasons other than water supply.

Other sectors, such as hydroelectric power generation, use much more water than other sectors but also recycle approximately 98% of the water that is used. In contrast, the irrigated farming sector uses a large share of the total water supply while returning only approximately 68% of the water back to the land; the rest is lost to evaporation.

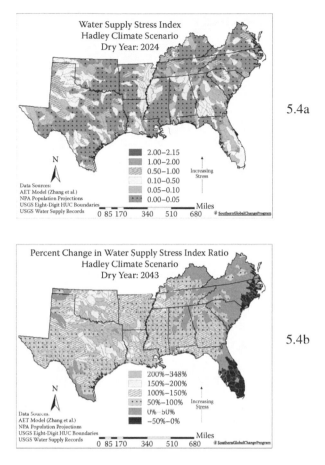

5.4a

5.4b

FIGURE 5.4 Population change impacts on the southeastern U.S. water supply stress (WASSI) during a dry year in 2024 (a) and a dry year in 2043 (b). Change in the water supply stress ratio (WASSIR) between 1990 and the two dry years of 2043 (c) and 2024 (d).

5.3.3 Ground Water Supply Controls on the WASSI

The loss of the groundwater resource can have severe implications for the WASSI in some areas where groundwater represents a major source of the water supply. By the 2040s, it is likely that areas with limited aquifer reserves and heavy groundwater use will begin to run out of groundwater. In our study, we expected loss of groundwater to have a severe impact on the WASSI during the dry year of 2043 (Figure 5.5a). It was somewhat surprising that even during the projected wet year of 2045, severe water stress would occur over much of the region without groundwater (Figure 5.5b). The water supply stress index ratio (WASSIR) increased by 232% between the base-line year of 1990 and the dry, no groundwater year of 2043 (Figure 5.5c). Even for the wet year/no groundwater scenario of 2045, the WASSIR difference between 1990 and the dry years without groundwater (i.e., 2024 and 2043) was 119% (Figure 5.5d).

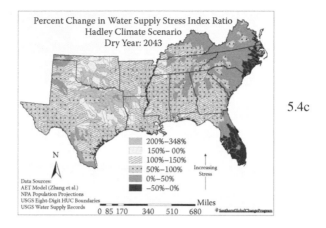

5.4c

5.4d

FIGURE 5.4 (Continued)

Land use planners should carefully review the implication of groundwater loss on local economies and consider converting land area from higher to lower water-consuming practices (e.g., reduce irrigated acreage) or the use of water transport systems to replace exhausted aquifers. Even the most optimistic estimates of climate change and increased precipitation will not likely alleviate future water stress, should aquifers run dry.

5.4 SUMMARY

This paper explored the likely impacts of climate, population, and groundwater supply on interannual water supply stress across the southeastern United States during the next 40 years. We found that predicted climate variability will have the largest impact on the water supply stress. However, the current WASSI model does not carry water reserves or deficits from one year to the next. There is no drawdown of water reservoir capacity as would occur during a prolonged drought. Similarly, there is no

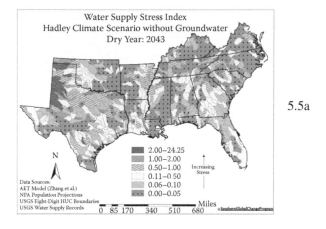

5.5a

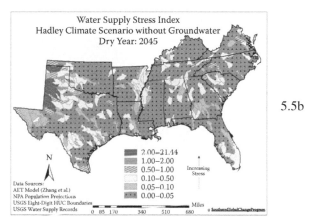

5.5b

FIGURE 5.5 Climate change impacts on the southeastern U.S. water supply stress (WASSI) during the dry year 2043 if no groundwater were available (a) and the wet year of 2045 with no available groundwater (b). Change in the water supply stress ratio (WASSIR) between 1990 and the dry year of 2043 if no groundwater were available (c) and the wet year of 2045 with no available groundwater (d).

reserve water capacity to compensate for water shortfalls during a drought. Therefore, the current WASSI model could overestimate the impact of short-term droughts because water deficits could be offset by water reservoirs.

Watersheds receiving limited precipitation and with a heavy dependence on groundwater will be the most susceptible to chronic and potentially permanent water shortages. Water use managers should expect even more stress in large population centers. Less-populated areas that had little water shortage problems in the past may also face water stress issues under changes in global and regional climate. However, future climate change–induced precipitation patterns remain uncertain, especially in the eastern United States, and thus realistically predicting future water stress remains a challenge.

5.5c

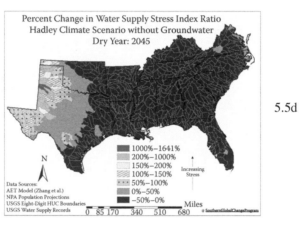

5.5d

FIGURE 5.5 (Continued)

ACKNOWLEDGMENTS

Funding for this work was provided by the U.S. Department of Agriculture Forest Service Southern Global Change Program in Raleigh, North Carolina. The authors thank Corey Bunch for programming support.

REFERENCES

Brown, R. A., N. J. Rosenberg, and R. C. Izarraulde. 1999. *Responses of U.S. regional water resources to CO₂-fertilization and Hadley center climate model projections of greenhouse-forced climate change: A continental scale simulation using the HUMUS model.* Richland, WA: Pacific Northwest National Laboratory.
Kittel, T. G. F., J. A. Royle, C. Daly, N. A. Rosenbloom, W. P. Gibson, H. H. Fisher, D. S. Schimel, L. M. Berliner, and VEMAP2 Participants. 1997. A gridded historical (1895–1993) bioclimate dataset for the conterminous United States. In *Proceedings of the 10th Conference on Applied Climatology.* Boston, MA: American Meteorological Society, 219–222.

Lu, J., G. Sun, D. M. Amatya, and S. G. McNulty. 2003. Modeling actual evapotranspiration from forested watersheds across the southeastern United States. *Journal of American Water Resources Association* 39(4):886–896.

NPA Data Services, Inc. 1999. Economic databases—mid-range growth projections 1967–2050. Regional Economic Projections Series, Arlington, VA: NPA Data Services.

Solley, W. B., R. R. Pierce, and H. A. Perlman. 1993. *Estimated use of water in the United States in 1990.* USGS Circular 1081. Reston, VA: U.S. Geological Survey.

Solley, W. B., R. R. Pierce, and H. A. Perlman. 1998. *Estimated use of water in the United States in 1995.* USGS Water Resources of the United States. USGS Circular 1200. Reston, VA: U.S. Geological Survey.

Sun, G., S. G. McNulty, J. Lu, D. M. Amatya, Y. Liang, and R. K. Kolka. 2005. Regional annual water yield from forest lands and its response to potential deforestation across the southeastern United States. *Journal of Hydrology,* 308:258–268.

U.S. Census Bureau. 2002. Perennial Census Data. Retrieved April, 15, 2002, from http://www.census.gov.

U.S. Geological Survey, Water Resources Division. 1994. *Hydrologic unit maps of the coterminous United States: 1:250,000 scale, 8-digit hydrologic unit codes and polygons.* Reston, VA: U.S. Geological Survey.

6 Study on the Intra-Annual Distribution Characteristics of the Water Budget in the Hilly Region of Red Soil in Northeast Jiangxi Province, China

Junfeng Dai, Jiazhou Chen, Yuanlai Cui, and Yuanqiu He

6.1 INTRODUCTION

In northeast Jiangxi Province, water balance is important for water resource utilization and agricultural regionalization. In this region, the red soil is affected by the subtropical monsoon climate in which rainfall is abundant, but unevenly distributed (Chen and Zhang 2002). The high intensity of rainfall leads to water loss during the rainy season. In addition, the time difference between the distribution of the rainfall and its evaporation causes seasonal drought. Studying the characteristics of the intra-annual distribution of the water budget can help discover the occurrence rules of seasonal drought in the region and assist in adopting measures to increase water use efficiency (Huang et al. 2004).

Existing literature has examined aspects of the water problem in red soil areas. Yao (1996) described the water problem including seasonal drought and water storage capacity in a red soil area. Wang et al. (1996) analyzed the variations in rainfall, evaporation, and water storage and supply in low, hilly red soil areas. Xie et al. (2000) studied the status of water resources in 1,000-m^2 area using the runoff plot, neutron probe, and tensiometer. However, the water budget in the hilly region of red soil has rarely been investigated systematically. It is difficult to obtain or measure percolation, evapotranspiration, and runoff accurately in the large- to medium-scale basin, and even in the small-scale watershed. Therefore, water budget in a scale larger than a field is not easily studied using traditional methods. Model simulation is a viable option for calculating the output of the water budget at a variety of scales. However, such studies have rarely been undertaken in the red soil areas of China. SWAT (Soil and Water Assessment Tool) (Arnold et al. 1998) is a distributed hydrological model based on physical mechanisms. Although SWAT is designed to predict the impact of management on water, sediment, and agricultural chemical yields in large ungauged basins, it is also valid in small catchments (Chanasyk et al. 2003, Mapfumo et al. 2004).

In this study, characteristics of intra-annual distribution of water balance are investigated using runoff catchments and the SWAT99.2 model in order to provide appropriate directions for effective management and use of water resources in the hilly region of red soil.

6.2 BRIEF DESCRIPTION OF MODEL

SWAT is a distributed and continuous time model developed by Dr. Jeff Arnold for the U.S. Department of Agriculture (USDA), Agricultural Research Service. It has undergone several versions with different interfaces. In this article, the results of water budget are calculated using the Windows interface for SWAT99.2.

The SWAT model consists of three major components: sub-basin, reservoir routing, and channel routing. The sub-basin component consists of eight main subcomponents defined as hydrology, weather, sedimentation, soil temperature, crop growth, nutrients, agricultural management, and pesticides. The following is a brief description of the main hydrology subcomponent. For a complete description, see Arnold et al. (1999).

SWAT simulates surface runoff volumes using daily rainfall amounts. Surface runoff volume is computed using a modification of the SCS (Soil Conservation Service) curve number method (USDA Soil Conservation Service 1972), and canopy storage is taken into account in the surface runoff calculations. The percolation component of SWAT uses a storage routing technique to predict flow through each soil layer in the root zone. Percolation occurs when the field capacity of a soil layer is exceeded and the layer below is not saturated. The flow rate is governed by the saturated conductivity of the soil layer. The model computes evaporation from soils and plants separately as described by Ritchie (1972). Potential soil water evaporation is estimated as a function of potential evapotranspiration and leaf area index. Actual soil water evaporation is estimated by using exponential functions of soil depth and water content. Plant transpiration is simulated as a linear function of potential evapotranspiration and leaf area index. The model offers three options for estimating potential evapotranspiration: Hargreaves, Priestley-Taylor, and Penman-Monteith. SWAT utilizes a simplification of the EPIC crop model (Williams et al. 1984) to simulate all types of land covers. The model is able to differentiate between annual and perennial plants.

6.3 SITE SELECTION AND MODEL CALIBRATION

6.3.1 APPLICATION SITE

The study area is located in the artificial forest and grass catchments of the Experiment Station of Red Earth Ecology, Chinese Academy of Science, Yingtan, Jiangxi Province (Figure 6.1).

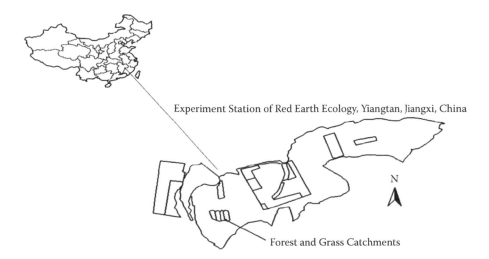

Experiment Station of Red Earth Ecology, Yiangtan, Jiangxi, China

N

Forest and Grass Catchments

FIGURE 6.1   Schematic map of the study area in China.

The catchments are at an altitude of 48 to 54 meters and the groundwater table is low. The catchments were established in 1988 and were isolated by concrete 5 cm wide and 1 m deep. The four catchments are relatively dependent. Consequently, each catchment is considered a watershed in the model and is not partitioned into sub-basins. Water budget in each catchment is calculated respectively.

An observation house with a runoff pool was constructed at the exit of every catchment, and a water level gauge was installed in the runoff pool. Surface runoff of catchments could be calculated through the continuous recording of the water level change.

The details of the four catchments are as follows: (1) Natural grass catchment: the vegetation recovered naturally after the original *Pinus massoniana* was cut; (2) Evergreen broad-leaved forest catchment: *Schima Superba* and *C. fissa* were planted after the *Pinus massoniana* was cut; (3) Mixed forest catchment: *Q. chenil* and *Lespedeza bicolor* were planted in the original *Pinus massonianas*; (4) Coniferous forest catchment: the original *Pinus massoniana* forestland remains.

6.3.2 Data Preparation

The model inputs require climate, soil, land use, and vegetation data. There is a meteorological observation station near the forest and grass catchments. Daily precipitation and maximum and minimum air temperature observations from 2000 to 2001 are available, and they are input directly into the SWAT model. In the SWAT99.2 model with Windows interface, daily solar radiation, wind speed, and relative humidity are generated from average monthly values based on the historical statistical data. These statistics are generated using a weather generator module in SWAT. Thirteen years of observational data have been used to generate these statistics.

The soil type of the catchments is red soil derived from the quaternary red clay. The properties of each soil layer are given in Table 6.1.

TABLE 6.1

Soil properties of forest and grass catchments.

Soil depth (mm)	Parameter	G	Fb	Fm	Fc
200.00	Bulk density(g/cm³)	1.45	1.29	1.31	1.42
	Organic carbon content (%)	0.55	0. 60	0.58	0.59
	Saturated conductivity (mm/h)	21.10	35.13	32.12	27.92
	Available water capacity (mm/mm)	0.09	0.10	0.09	0.09
	Clay content (%)	39.95	41.20	40.20	39.05
	Silt content (%)	36.15	39.70	36.45	32.15
	Sand content (%)	23.9	19.10	23.35	28.80
600.00	Bulk density (g/cm³)	1.27	1.28	1.26	1.24
	Organic carbon content (%)	0.30	0. 39	0.38	0.37
	Saturated conductivity (mm/h)	4.48	8.78	6.78	3.80
	Available water capacity (mm/mm)	0.10	0.08	0.09	0.09
	Clay content (%)	48.45	50.95	49.75	50.05
	Silt content (%)	37.55	37.55	37.30	33.85
	Sand content (%)	14.00	11.50	12.95	15.80
2000.00	Bulk density (g/cm³)	1.25	1.27	1.25	1.23
	Organic carbon content (%)	0.20	0. 23	0.23	0.23
	Saturated conductivity (mm/h)	1.24	2.76	1.72	1.08
	Available water capacity (mm/mm)	0.08	0.08	0.07	0.08
	Clay content (%)	52.65	52.05	52.10	53.10
	Silt content (%)	33.70	37.00	36.20	34.00
	Sand content (%)	13.65	10.95	11.70	12.90

Note: G: natural grass; Fb: broad-leaved forest; Fm: mixed forest; Fc: coniferous forest.

Watershed attributes and vegetation characteristics of catchments are gained with investigation (Table 6.2). For the forest and grass catchments, a wet and a dry year (2000 and 2001) of daily runoff are obtained from the Experiment Station of Red Earth Ecology to calibrate and validate the model.

6.3.3 MODEL CALIBRATION AND VALIDATION

The SWAT model is calibrated and validated against the observed runoff. In this study, the SCS curve number (CN_2) and soil evaporation compensation factor (ESCO) are adjusted until surface runoff is acceptable. CN_2 is a function of the soil's permeability, land use, and antecedent soil water conditions. The parameter calibration of the model is displayed in Table 6.3.

The correlation coefficient (R^2) and Nash-Sutcliffe coefficient (E_{ns}) (Saleh et al. 2000) are used to evaluate the variance between observed and simulated values. Comparing simulated daily and monthly surface runoff to observations, model efficiency is achieved (Table 6.4). It shows that the simulation model can be used as an analytical tool for calculating the hydrological cycle in a hilly region of red soil.

TABLE 6.2

Watershed attributes of forest and grass catchments.

Parameter	G	Fb	Fm	Fc
Area (m²)	2672.2	3162.0	2483.6	3134.9
Slope length (m)	96.0	90.0	98.1	93.0
Slope steepness (m/m)	0.0573	0.0556	0.0535	0.0484
Surface litter (kg/hm²)	172.1	5360.0	3294.4	1492.5
Land cover	Perennial	Trees	Trees	Trees
Ground cover	>75%	>75%	>75%	>75%

TABLE 6.3

Parameter calibration of model.

	G		Fb		Fm		Fc	
Parameter	Original	Calibrated	Original	Calibrated	Original	Calibrated	Original	Calibrated
CN_2	71	68	65	63	65	64	65	65
ESCO	0.95	0.80	0.95	0.70	0.95	0.75	0.95	0.75

TABLE 6.4

Evaluation of simulation efficiency.

Year		G		Fb		Fm		Fc	
		Daily	Monthly	Daily	Monthly	Daily	Monthly	Daily	Monthly
2000	R^2	0.91	0.91	0.92	0.93	0.89	0.90	0.89	0.92
	E_{ns}	0.63	0.74	0.72	0.76	0.60	0.73	0.61	0.72
2001	R^2	0.86	0.90	0.73	0.85	0.78	0.93	0.80	0.89
	E_{ns}	0.74	0.73	0.66	0.75	0.65	0.74	0.67	0.73

6.4 RESULTS AND DISCUSSION

6.4.1 PRECIPITATION

The daily rainfall record over a period of two years (Figure 6.2) indicates that the rainfall in the red soil area is uneven in its intra-annual distribution.

The annual rainfall is heavy in the red soil area with 1,912.1 mm falling in 2000 and 1,482.4 mm for 2001. The daily rainfall record also shows that the heaviest rains fall from April to July. The April to June 2000 rainfall amounts account for about

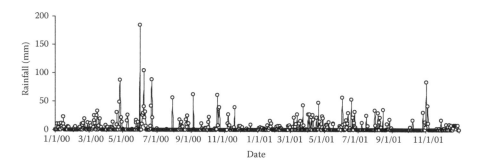

FIGURE 6.2   Daily rainfall in the study area.

50.33% of annual total rainfall. High rainfall intensity is frequently observed during the rainy season (April to June) in the red soil areas. Over the two-year period, there were 10 days where the daily rainfall exceeded 50 mm, totaling 841.5 mm and representing 24.8% of the two-year total. In addition, 47.8% of the two-year rainfall occurs within a 38-day period, for a rain intensity of 42.7 mm d^{-1}. The rainfall pattern in the red soil area is hazardous to water resource utilization and crop growth.

6.4.2 SURFACE RUNOFF

Daily observed and simulated surface runoffs are displayed in Figure 6.3. Both observed data and simulation results indicate that surface runoff scarcely takes place when the daily rainfall is less than 10 mm. The surface runoff can be observed when the daily rainfall is above 10–15 mm. The results show that temporal change of rainfall led to the variation of the surface runoff. Compared with the 2000 data, rainfall in 2001 decreased 22.74%, and the surface runoff of the four catchments was reduced between 62.11% and 74.04%. Furthermore, the surface runoff often occurs in the rainy season. The concentrated rainfall pattern induces heavy runoff. For instance, the rate of three months (April to June) surface runoff of G, Fb, Fm and Fc is 69.21%, 80.02%, 73.72%, and 69.41%, respectively, of the annual 2000 amount. Also, the surface runoff scarcely occurs in January, February, and December due to the shortage of rainfall.

In addition, the average annual runoff coefficient of the forest and grass catchments is 0.14, which is different from the value of 0.1 in long-term cultivated red soils (Yang et al. 1993), and which also differs from the value ranging from 0.55 to 0.61 in the uncultivated or newly cultivated uplands of red soils (Ju and Wu 1995). These reveal that runoff in the red soil uplands changes remarkably with different types of utilization and vegetation.

6.4.3 EVAPOTRANSPIRATION

Potential evapotranspiration is estimated using the Penman–Monteith equation in this article. Compared with the observed daily water surface evaporation, the calculated daily evapotranspiration in the forest and grass catchments is reasonable (Figure 6.4). Evapotranspiration mainly concentrates from April to August, especially between June and August, because of the ample soil water and high temperature. Between

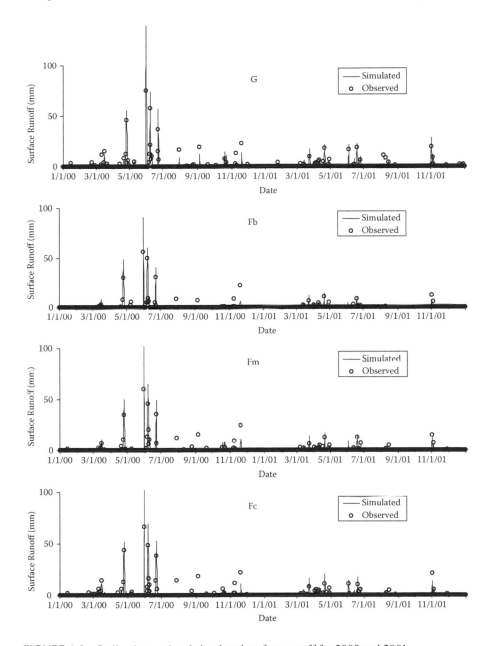

FIGURE 6.3   Daily observed and simulated surface runoff for 2000 and 2001.

June and August 2000, the evapotranspiration of the catchments accounted for 45.69 to 50.87% of the annual total.

Rainfall decreases sharply after August, but the actual evapotranspiration loss is still large. The time period of maximum evapotranspiration is not synchronous with that of maximum rainfall, which is the essential intra-annual distribution of the water budget in the red soil areas.

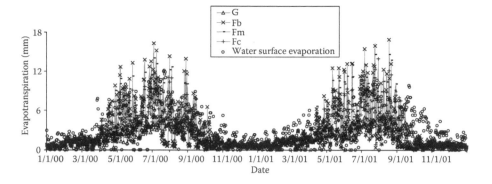

FIGURE 6.4   Simulated daily evapotranspirations in the forest and grass catchments.

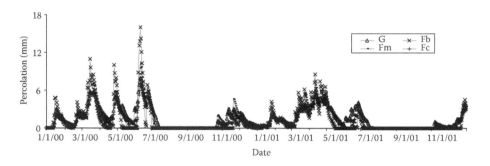

FIGURE 6.5   Simulated daily percolations from the root zone in forest and grass catchments.

In addition, the annual difference in evapotranspiration of forest and grass eco-systems is small. The results show that the evapotranspiration decreases less with rainfall than with surface runoff with rainfall. For instance, comparing 2001 data with that of 2000, rainfall decreases 22.47%, runoff of broad-leaved forest reduces 75.30% accordingly, but evapotranspiration only decreases 7.32%.

6.4.4 Percolation from the Bottom of the Root Zone

Figure 6.5 shows that percolation from the root zone of forestland and grassland often occurs in the rainy season. The excessive rainfall usually induces large daily percolation. The rate of four months' (March to June) percolation is 67.00 to 79.35% of the annual rainfall over a period of two years. Percolation from the root zone scarcely occurs the rest of the year (July to October) due to insufficient rainfall and strong evapotranspiration. As a consequence of unequal rainfall, the rate of 2001 percolation of catchments is between 75.09 and 81.54% of the 2000 rate. In addition, the permeability of forest and grass catchments in red soil is better than that in barren land in this area. This reveals that vegetation recovery in the red soil area can assist soil infiltration and water resource utilization.

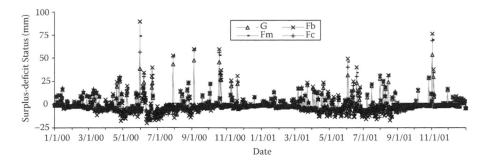

FIGURE 6.6   Simulated daily water surplus-deficit status of forest and grass catchments.

6.4.5 WATER SURPLUS-DEFICIT STATUS

In the study area, rainfall is the water input, and water outputs include surface runoff, percolation, and evapotranspiration. The water surplus-deficit status in the catchments is shown in Figure 6.6. Water deficits mainly occur from June to September, but in other months, water input is generally larger than water output. The results indicate that annual change of soil water storage accounts for about 1.0% of rainfall.

From February to April, increasing rainfall and weak evaporation potential contribute to the increase in soil water storage, although a water deficit can exist. From May to September, the soil water storage evinces a gradual decreasing trend. Especially from July to September, the forestlands have a serious deficit of soil water due to lack of rainfall and strong evaporation potential, and the depleted water is recharged by the soil water storage. Furthermore, in the autumn sowing period, a serious water shortage also occurs on farmland. This conflict is detrimental to water utilization and crop growth. From October to the following February, the soil water storage gradually reaches the proper status. As a whole, the annual change of soil water storage is quite small. This indicates that the water budget in the hilly region of red soil has a dynamic balance.

The results also show that the evapotranspiration in forest and grass catchments accounts for a majority of the water output in the red soil area. The proportion between annual evapotranspiration and rainfall is above 0.5 in forestlands, and approximately 0.4 in the grass catchment. Furthermore, in the forest and grass catchments where vegetation has grown for twelve years or more, percolation from the root zone is bigger than surface runoff, and the condition seems more remarkable in the wet year.

6.5 CONCLUSIONS

The main characteristics of the water budget in a hilly region of red soil investigated with runoff catchments and model simulation revealed the following information.

The temporal distribution of rainfall is uneven in red soil areas. High rainfall intensity is frequently observed during the rainy season (April to June), which induces heavy runoff. Maximum evapotranspiration occurs from July to September, which is not synchronous with maximum rainfall. This conflict is detrimental to water utilization. In the forest and grasslands where vegetation has grown for twelve years or more, evapotranspiration is the largest water loss component, followed by percolation, and then surface runoff. This differs from the long-term cultivated lands, where percolation is the largest water loss component, and also from the uncultivated uplands, where runoff is the largest component.

The variation in weather conditions, especially rainfall, is the most important factor that results in the difference of water balance in the red soil area. Of the water outputs, the surface runoff and percolation are most affected by rainfall, followed by evapotranspiration, and the annual fluctuation of the soil water storage is very small.

According to the intra-annual distribution characteristics of the water budget in the red soil, important measures such as storing rainfall through the rainy season, constructing the irrigation works, agroforestry system or silvopastoral system techniques, and agricultural water-saving techniques should be taken to increase the water use efficiency and relieve the seasonal drought in the hilly red soil region.

This study also confirms that the modeling tool is an effective method for investigating the water balance of the hilly red soil region. The modeling tool is demonstrated in the forest and grass catchments in this research, but it could be adapted to the cropland catchments or large-scale watersheds in the red soil areas of China. The advantages of the modeling tool are not limited to its capability to simulate hydrological processes. Many management scenarios, such as fertilization, irrigation, water deficits, and population control, could also be analyzed.

ACKNOWLEDGMENTS

This study was financially supported by the National Natural Science Foundation of China (No. 40301019) and Chinese Ministry of Education Project (No. NCET-04-0664).

REFERENCES

Arnold, J. G., R. Srinivasan, R. S. Muttiah, and J. R. Williams. 1998. Large area hydrologic modeling and assessment. Part I, model development. *Journal of the American Water Resource Association* 34(1):73–89.

Arnold, J. G., J. R. Williams, R. Srinivasan, and K. W. King. 1999. Soil and water assessment tool theoretical documentation (version 99.2). http://www.brc.tamus.edu/swat.

Chanasyk, D. S., E. Mapfumo, and W. Willms. 2003. Quantification and simulation of surface runoff from fescue grassland watersheds. *Agricultural Water Management*, no. 59: 137–153.

Chen, Z. F., and X. X. Zhang. 2002. Effect of seasonal drought on forestry and fruit industry at red soil region in southern China [in Chinese]. *Agro-environmental Protection* 21(3):241–244.

Huang, D. Y., K. L. Wang, M. Huang, H. S. Chen, J. S. Wu, G. P. Zhang, and T. B. Peng. 2004. Seasonal drought problems in the hilly, red soil region of the middle subtropical zone of China [in Chinese]. *Acta Ecologica Sinica* 24(11): 2516–2523.

Ju, Z. H., and S. Z. Wu. 1995. The soil water balance conditions and improvement ways in central Jiangxi Province. In *Agricultural integrated development and countermeasure in red and yellow soil area in China* [in Chinese], ed. Y. S. Yang and L. Q. Xin. Beijing: China Agricultural Science and Technology Publishing House, 143–147.

Mapfumo, E., D. S. Chanasyk, and W. D. Willms. 2004. Simulating daily soil water under foothills fescue grazing with the soil and water assessment tool model (Alberta, Canada). *Hydrological Processes* 8: 2787–2800.

Ritchie, J. T. 1972. A model for predicting evaporation from a row crop with incomplete cover. *Water Resource Research* 8:1204–1213.

Saleh, A., J. G. Arnold, P. W. Gassman, L. M. Hauck, W. D. Rosenthal, J. R. Williams, and A. M. S. McFarland. 2000. Application of SWAT for the upper north Bosque River watershed. *Transactions of the ASAE* 43(5):1077–1087.

U.S. Department of Agriculture, Soil Conservation Service. 1972. *National Engineering Handbook*. Section 4, Hydrology. Washington, DC: U.S. Department of Agriculture, chaps. 4–10.

Wang, M. Z., J. B. Zhang, and C. S. Zhao. 1996. Study on temporal and spatial variations of water resources and its comprehensive utilization in low hilly red soil area [in Chinese]. *Acta Agriculturae Jiangxi* 8(1):47–58.

Williams, J. R., C. A. Jones, and P. T. Dyke. 1984. A modeling approach to determining the relationship between erosion and soil productivity. *Transactions of ASAE* 27.129–144.

Xie, X. L., K. R. Wang, and W. J. Zhou. 2000. Status and management of water resources in red soil regions and hilly-sloping lands [in Chinese]. *Journal of Mountain Science* 18(4):336–340.

Yang, Y. S., D. M. Shi, X. X. Lu, and Y. Liang. 1993. Analysis of sediment delivered from eroded sloping land in quaternary red region. In *Research on red soil ecosystem* [in Chinese], ed. M. Z. Wang, T. L. Zhang, and Y. Q. He. Nanchang: Jiangxi Science and Technology Publishing House, 330–344.

Yao, X. L. 1996. Water problem of red soil and its management [in Chinese]. *Acta Pedologica Sinica* 33(1): 13–20.

7 Forest and Water Relations

Hydrologic Implications of Forestation Campaigns in China

Ge Sun, Guoyi Zhou, Zhiqiang Zhang,
Xiaohua Wei, Steven G. McNulty, and James Vose

7.1 INTRODUCTION

Forest inventory records indicate that the forested area in China fell from 102 million hectares in 1949 to approximately 95 million hectares in 1980 due to accelerated population growth, industrialization, and resource mismanagement during that period (Fang et al. 2001, Liu and Diamond 2005). Consequently about 38% of China's land mass is considered badly eroded (Zhang et al. 2000) due to deforestation and rapid urbanization (Liu et al. 2005). However, forest coverage is recovering (Liu and Diamond 2005), and China now has the largest area of forest plantations in the world, accounting for approximately 45 million ha, which is one fourth of the world total (Food and Agriculture Organization [FAO] 2004, http://www.fao.org/) (Figure 7.1). A new forest policy, called the Natural Forest Conservation Program (NFCP), was adopted after the severe floods of 1998 (Zhang et al. 2000). This policy's objectives include restoring natural forests in ecologically sensitive areas such as the headwaters of several large rivers, including the Yangtze River and the Yellow River, planting trees for soil and water protection, increasing timber production through forest plantations, banning excessive cutting, and maintaining the multiple use of forests. China's massive forestation plan (Program for Conversion of Cropland to Forests) aims to increase forested areas by 440,000 km^2 or 5% of its landmass in the next 10 years (Lei 2002). This includes 14.66 million ha of soil erosion–prone croplands that will be converted to forests and 17.33 million ha of barren land that will be revegetated during the next ten years.

Plot-scale studies in China have documented that reforestation and forestation can reduce soil erosion and sediment transport (Zhou and Wei 2002) and enhance carbon sequestration (Fang et al. 2001). However, surprisingly, few rigorous long-term studies in China have examined the relationship between water quantity and quality and forestation activities at watershed and regional scales. The impacts of the massive forestation efforts described above on watershed hydrology and water resources have not been as well studied in China or in the forest hydrology community. Scientific

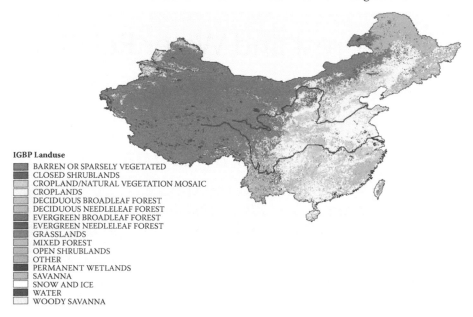

IGBP Landuse
- BARREN OR SPARSELY VEGETATED
- CLOSED SHRUBLANDS
- CROPLAND/NATURAL VEGETATION MOSAIC
- CROPLANDS
- DECIDUOUS BROADLEAF FOREST
- DECIDUOUS NEEDLELEAF FOREST
- EVERGREEN BROADLEAF FOREST
- EVERGREEN NEEDLELEAF FOREST
- GRASSLANDS
- MIXED FOREST
- OPEN SHRUBLANDS
- OTHER
- PERMANENT WETLANDS
- SAVANNA
- SNOW AND ICE
- WATER
- WOODY SAVANNA

FIGURE 7.1   Land cover of China as classified by the IGBP (International Geosphere-Biosphere Programme) system. (**See color insert after p. 162.**) The majority of the forest-lands are located in the hilly remote southwestern and northeastern regions.

debates on the hydrologic role of forests intensified when floods struck, such as in 1981 and 1998.

The objectives of this paper are: (1) to synthesize existing worldwide literature on the relations between forestation and watershed hydrology, (2) to identify factors affecting hydrologic responses to forestation, (3) to discuss the potential hydrologic consequences of large-scale vegetation-based watershed restoration efforts in China, and (4) to recommend future forest hydrologic research activities to guide watershed ecological restoration campaigns.

7.2 FORESTS AND WATERSHED HYDROLOGY: EXPERIMENTAL EVIDENCE AROUND THE WORLD

Many *paired watershed* manipulation studies addressing forest–water relations have been conducted in the past 100 years around the world, published in English (Hibbert 1967, Bosch and Hewlett 1982, Ffolliott and Guertin 1987, Whitehead and Robinson 1993, Stednick 1996, Sahin and Hall 1996, Scott et al. 2005, Brown et al. 2005, Farley et al. 2005) as well as in Chinese (Wang and Zhang 1998, Li 2001, Liu and Zeng 2002, Zhang et al. 2004, Wei et al. 2005b). Key research results in the international literature and in China are listed in Table 7.1 to facilitate the discussion and for future reference. Below are examples of the highlights of studies on the effects of forestation on watershed hydrology grouped by continent. Watershed hydrologic impact studies are discussed in terms of changes in total annual water yield, storm-flow rates and volume, and baseflow rates and volumes.

TABLE 7.1

Key publications on forest–water relations.

References	Region Ecosystems	Key Findings
Bosch and Hewlett 1982	Worldwide; all ecosystems	Annual evapotranspiration decreases with vegetation removal
Andreassian 2004	Worldwide, all ecosystems	Deforestation (reforestation) increases (decreases) water yield; the variability can be explained by differences in climate, soil, and vegetation characteristics
Jackson et al. 2005	Worldwide, all ecosystems	Plantations reduce stream flow, and increases soil salinization and acidification
Ice and Stednick 2004	United States; all type of forests	Deforestation increases water yield
Beschta et al. 2000	Western Cascade of Oregon, United States	Forest harvesting increases small-sized peak flows; not likely to cause peak flow increases in large basins
Scott et al. 2005	South Africa and Tropics; forest plantations	Converting grasslands or reforesting degraded lands with plantations reduce base flow and water yield; little impacts on peak flows
Robison et al. 2003	European forest ecosystems	Similar results to North America; forests play small role in water resource management for floods and droughts
Brown et al. 2005	Australia and worldwide; all forest ecosystems	Variable hydrologic recovery time for deforestation and reforestation, which mainly impact base flows
Ma 1987	Sub-alpine, southwestern China	Water yield increased after forest harvesting treatment
Liu and Zhong 1978	Loess Plateau, China	Water yield was lower in forested watersheds
Wei at al. 2005a, 2005b	China	Contradictory data on forest–water relations
Sun et al. 2006	China	Simulated water yield reduction following reforestation most significant in northern regions

7.2.1 NORTH AMERICA

North America contains a diverse mixture of forest ecosystems, from boreal forests in Canada, in which snow often dominates the hydrologic processes, to semiarid-arid shrub lands in the southwestern United States where water stress is common. Long-term experimental stations, including the Coweeta Hydrologic Laboratory, Hubbard Brooks, and Andrews Experimental Forests in the United States (Figure 7.2), and the Turkey Lakes Watershed Study in Canada, were designed to answer watershed management questions specifically related to water quantity and quality. Many of the experimental watersheds have provided over 50 years of continuous forest hydrologic data. Much of our current understanding of modern forest hydrological and ecosystem processes has been derived from these watersheds.

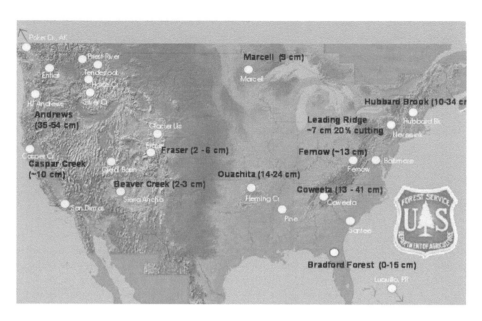

FIGURE 7.2 First-year water yield response to deforestation (clear-cut) varies across the physiographic gradient in the United States.

Experimental results in the United States have been synthesized by Hibbert (1967), Bosch and Hewlett (1982), in a special issue of the *American Water Resource Bulletin* published in 1983, by Post and Jones (2001), and more recently in a book by Ice and Stednick (2004). Canadian forest hydrology research activities were summarized by Buttle et al. (2000, 2005). Long-term empirical data across the physiographic gradients in the United States suggest diverse watershed hydrologic response to forest removal (Figure 7.2). For example, a 46-year paired watershed study at the Coweeta Hydrologic Laboratory in a humid subtropical climate with deep soils shows that repeated cutting of mountain forests can increase streamflow by 200 to 400 mm per year. The hydrologic effects lasted more than 20 years (Swank et al. 1988). Streamflow decreased with the regeneration and regrowth of the deciduous forests. A second cutting returned to pretreatment water yield faster than the first cutting cycle (Figure 7.3). Hydrologic responses differ across landscapes (i.e., upland vs. wetlands) and climatic conditions (Sun et al. 2004, Sun et al. 2005). Several field and modeling studies in the southeastern United States showed that forest management impacts on water yield were most pronounced during dry periods when trees that have deep roots can use moisture in subsurface soil layers (Trimble and Weirich 1987, Sun et al. 1998, Burt and Swank 2002). The effects of forests on annual water yield are propagated through their influence on baseflow. North American literature on forestry impacts on floods is more contentious (Jones and Grant 1996, Thomas and Megahan 1998, Beschta et al. 2000) than on annual water yield and baseflow. However, it is generally accepted that forest management affects small to moderate peak flow rates, but has little impact on large floods (Hewlett 1982, Burt and Swank 2002).

Reviews of Canadian forest hydrology by Buttle et al. (2000, 2005) concluded that watershed-scale studies to evaluate the hydrologic effects of large-scale forest

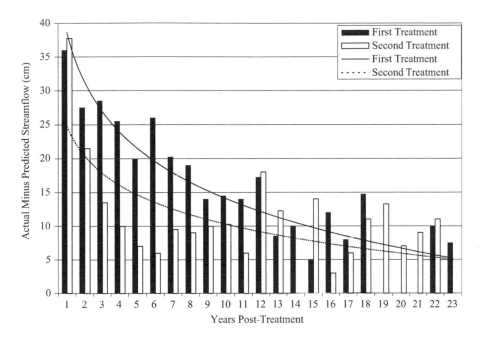

FIGURE 7.3 Annual streamflow responses to repeated harvesting of mixed hardwood forest on watershed 13 at the Coweeta Hydrologic Laboratory located in the southern Appalachian Mountains. (Adapted from Swank et al. 1988, Streamflow changes associated with forest cutting, species conversion, and natural disturbances. In *Ecological studies*. Vol. 66, *Forest hydrology and ecology at Coweeta*, ed. W. T. Swank and D. A. Crossley Jr. New York: Springer-Verlag, 297–312.)

removal for managing recent fire and insect disturbances are lacking in Canada. Limited watershed manipulation studies suggest that drainage through ditching increased baseflow, but not peak flow in a Quebec peat land. Peak flow rates were not affected significantly in a watershed in New Brunswick with a 23.4% forest removal. Buttle et al. (2005) cautioned that importation and direct application of results from other regions in the United States to Canada may not be appropriate due to the unique geological (i.e., glacier vs. nonglacier) and climatic conditions (e.g., snow dominated vs. rain dominated), and because the treatment methods used in the 1960s and 1970s by U.S. researchers are no longer in use.

7.2.2 EUROPE

Forest is a major land cover type in Europe, and recent droughts and floods have attracted new interest in the role of forests in influencing river flow regimes. In a synthesis study across the European continent, Robinson et al. (2003) found that conifer plantations on poorly drained soils in northwestern Europe and eucalyptus in southern Europe may have marked local impacts on water yield similar to those reported in North America. However, changes of forest cover will not likely have great effect on extreme flows (i.e., floods and droughts) at the regional scale.

Robinson et al. (2003) stress the dilution effects of water flow for large basins, and conclude that forests have a relatively small role in managing risks of large-scale floods and droughts across the region.

7.2.3 SOUTH AFRICA AND THE TROPICS

It is estimated that 40 to 50 million ha of forest plantations grow in the tropics and warmer subtropics with an additional 2 to 3 million ha planted every year (Scott et al. 2005, Farley et al. 2005). The hydrologic impacts of forestation are more pronounced in this region due to the high water uptake by tropical trees. For example, some studies have recorded water yield increase of 80 to 90 mm per year per 10% forest removal (Bruijnzeel 1996, Bruijnzeel 2004). The response is much higher than the 25 to 60 mm per year range in the classic synthesis paper by Bosch and Hewlett (1982). A review of the literature on the humid tropical regions suggests the prospects of enhanced rainfall and augmented baseflow from reforestation are generally poor in most areas (Scott et al. 2005). A long-term (since the 1930s) paired watershed study for converting natural grasslands to forests with negative or exotic tree species in South Africa provided a comprehensive understanding of the hydrologic effects of forestation (Smith and Scott 1997, Scott et al. 1998). This study found that annual streamflow reduction rates increased over time following a similar sigmoidal pattern of tree growth. The highest flow reductions occurred when the plantations reached maturity. For every 10% level of planting, the reductions varied from 17 mm (or 10% per year) in a drier watershed to 67 mm (or 7% per year) for a wetter watershed. The low and high values are similar to those found in South India and Fiji respectively, and are within the range noted by Bosch and Hewlett (1982). This South Africa forestation study found that it took two years to have an appreciable reduction in streamflow after *Eucalyptus grandis* was planted over 97% of a native grassland watershed. However, it took eight years to have a clear streamflow impact after *Pinus patula* was planted over 86% of a native grassland watershed. The former reached the maximum streamflow reduction potential in about 15 years, while the latter did not reach the maximum reduction 25 years after planting. A recent update on this study reported that the reductions diminished after the plantations reached maturation, suggesting productive, vigorous growing forests use more water than mature or old, less vigorous growth forests (Scott et al. 2005). Finally, this long-term study concluded that forestation reduced total stream water yield, mostly baseflow, and can result in the complete loss of streamflow during the summer. Scott et al. (2005) postulated that the effect of forestation on streamflow decreased with storm size, and forestation had little effect on large storms when the soil conditions were not affected. Stormflows were mostly affected by soil water storage capacity and antecedent soil moisture conditions. Researchers in the tropics stressed the importance of differentiating *degraded lands* with bad soils versus *undisturbed good soils* that have very different soil hydrologic properties and processes when evaluating the effects of forestation on watershed hydrology (Bruijnzeel 2004, Scott et al. 2005). However, few definitive conclusions can be drawn from the literature on how forestation affects stormflows and baseflows. Few available studies suggest that revegetating degraded watersheds is not likely to augment baseflow and reduce stormflow volumes.

7.2.4 AUSTRALIA

Paired watershed manipulation studies in Australia produced a large amount of process-based information and useful models studying the effects of forestation on streamflow (Vertessy 1999, 2000; Zhang et al. 2001). Several Australian studies concluded that vigorous tree regrowth on cleared watersheds that were previously covered by old growth forests (e.g., mountain ash) resulted in decreased water yield due to increased evapotranspiration. Water yield from eucalyptus forests was found to be closely related to tree age (Cornish and Vertessy 2001, Vertessy et al. 2001). Vertessy and Bessard (1999) warned about the potential negative hydrologic effects (reduction of streamflow) of large-scale plantation expansion in Australia basins.

Andreassian (2004) and Brown et al. (2005) reviewed worldwide paired watershed experiments located in various geographic regions around the world. Highlights of the recent synthesis studies are summarized below with a focus on forestation effects.

The paired watershed experiments have crucial values in understanding the forest–water relationships. Existing paired watershed experiments are mostly designed for studying the effects of deforestation. Studies on reforestation are rare. Flow duration curve analysis methods provide insights on the seasonal effects of vegetation changes.

In general, deforestation increases annual water yield, and reforestation decreases it in proportion to vegetation cover change (Sun et al. 2006, Figure 4). Seasonal water yield response is variable (Brown et al. 2005), and is strongly influenced by precipitation patterns.

In general, deforestation increases flood volumes and peaks due to soil disturbances, but the effect is extremely variable. Limited studies on reforestation suggested that revegetation had minimal effect on small to moderate floods, and had no effect on flooding events.

Deforestation increases low flow (baseflow) and reforestation decreases it (Farley et al. 2005, Jackson et al. 2005).

7.3 DEBATE ON FOREST–WATER RELATIONS IN CHINA

Flooding and drought events cause huge economic losses each year in this heavily populated country. The Chinese people have long recognized the importance of forest and water to the environment and human societal development (Yu 1991). Because of the uncertainty of the relations between water resources and forests (Wei et al. 2005), great confusion and misconceptions regarding the hydrologic role of forests remain today (Zhou et al. 2001).

In the 1980s, studies on forest–water relations began to emerge in China (Ffolliott and Guertin 1987). Most of the studies have focused on the benefits of forests in retaining water for discharge during non-rainfall seasons (water redistribution) and on reducing floods during rainy seasons. Unfortunately, empirical observation and limited data on the environmental influences of forests, especially on hydrologic cycles, are often inconclusive and even contradictory (Wei et al. 2003, Wei et al. 2005) due to the highly diverse hydrologic processes caused by the large geographic and climatic variability in China.

Nevertheless, several well-cited studies have demonstrated the uncertainty and variability of potential hydrologic responses in China because of the large differences in climate and soil conditions. Liu and Zhong (1978) reported that forested watersheds on loess soils had a lower water yield amount (25 mm/yr) and a lower water yield/precipitation ratio than adjacent nonforest regions. This work was based on water balance data of several large basins in the upper reaches of the Yellow River in northwestern China. It was further estimated that forests in the Loess Plateau region may reduce annual streamflow by 37%. This study suggested that forested watersheds had higher total evapotranspiration, lower surface flow, but higher groundwater flow (baseflow). A three-year study in a small watershed in the middle reach of the Yellow River concluded that well-vegetated watersheds dominated by black locust (*Robinia pseudoacacia*) plantations and native pine species had over 100 mm per year higher evapotranspiration than the nonvegetated watersheds (Yang et al. 1999). Stormflow volume and peak flow rates were lower in the vegetated watersheds. The average annual precipitation was about 400 mm. Greater than 95% of precipitation evapotranspirated, and less than 5% precipitation became streamflow as infiltration-excess overland flow. The high tree density of plantations in the Loess Plateau region has resulted in low soil moisture in the rooting zone, which threatens the tree productivity and overall sustainability of the forestation efforts. A rare paired watershed experiment at a hardwoods forest site in northeastern China (annual precipitation = 700 to 800 mm) concluded that a 50% thinning caused total runoff to increase 26 to 31 mm per year (Ma 1993). However, several rather contradictory reports also exist. For example, Ma (1987) compared runoff between an old-growth fir forest watershed and a clear-cut watershed in the subalpine region of southwestern China, a tributary of the Yangtze River. This study was conducted in 1960, and found that water yield from the 331-ha forested watershed was much higher (709 mm/yr and a runoff ratio of 70.2%) than the 291-ha clear-cut watershed (276 mm/yr and a runoff ratio of 27.3%). In 1969, 60% of the forested watershed was harvested and water yield decreased by 380 mm per year. Detailed explanation of the causes of the hydrologic changes were not provided.

A comparison of streamflow from ten large basins (674 to 5,322 km²) in the Yangtze River showed that higher forest coverage generally had a higher runoff-to-rainfall ratio (>90%) (Ma 1987). Similar positive correlations between forests and water yield for large basins (>100 km²) were reported for northern China as cited in Wei et al. (2003). These findings corroborate Russian literature that suggests streamflow is generally higher for large forested basins (Wei et al. 2003). One unsubstantiated argument on the increase of streamflow from forests was that forest increased *fog drip* precipitation and that forests have lower evapotranspiration. Reports from studies in Russia on the forest–water relations had a large impact in China before the 1980s when access to Western literature was limited.

Wei et al. (2003) attributed the inconsistency of the studies described above to several reasons: (1) heterogeneous large basins have a large buffering capacity (e.g., wetlands) and may mask the forest cover effects, (2) inconsistent methods and measurement errors, and (3) differences in climate and watershed characteristics among the contrasting basins may obscure the forest cover effects. In fact, most of the watershed studies in China presented above did not follow the paired watershed principles, thus conclusions are subject to errors.

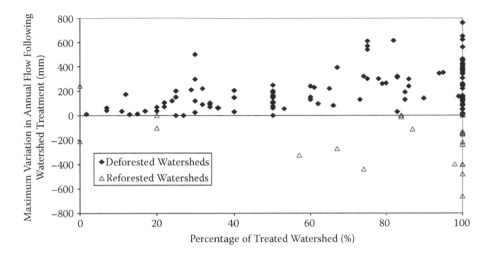

FIGURE 7.4 Worldwide review of paired watershed experiments on the stream flow response to deforestation and reforestation. (From V. Andreassian, Waters and forests: from historical controversy to scientific debate. *Journal of Hydrology* 291:1–27.)

Sun et al. (2006) examined the sensitivity of water yield response to forestation across China by employing a simple evapotranspiration model (equation 7.1) developed by Zhang et al. (2001) and a set of continental-scale databases including climate, topography, and vegetation (Sun et al. 2002). The Zhang et al. (2001) model was recently evaluated by Brown et al. (2005) using worldwide paired watershed studies. They found that the model is satisfactory at predicting the hydrologic effects of forestation of hardwoods and eucalyptus, but underestimates the effects for conifers. The model application study by Sun et al. (2006) concluded that forestation would have variable potential impacts across the diverse physiographic region (Figures 7.5 and 7.6). On average, the absolute values of reduction in water yield due to forestation ranged from approximately 50 mm per year in the drier northern region to about 300 mm per year to the southern humid region. This represents a 40% and 20% water yield reduction in the north and south, respectively. The predicted water yield reduction values reflect the climate (i.e., precipitation and potential evapotranspiration) controls on hydrologic responses to forestland cover changes. The predicted hydrologic responses are in the lower end of reported values when compared to the worldwide literature (Figure 7.4).

$$\Delta Q = ET_1 - ET_2 = -\left(\frac{1 + 2.0\dfrac{PET}{P}}{1 + 2.0\dfrac{PET}{P} + \dfrac{P}{PET}} - \frac{1 + 0.5\dfrac{PET}{P}}{1 + 0.5\dfrac{PET}{P} + \dfrac{P}{PET}} \right) \times P \quad (7.1)$$

where ΔQ = annual water yield change; ET_1, ET_2 = evapotranspiration of forest lands and grasslands, respectively; P = annual precipitation; PET = potential evapotranspiration calculated using Hamon's method as a function of monthly air temperature (Federer and Lash 1978).

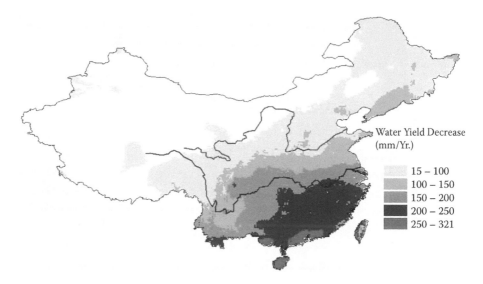

FIGURE 7.5 Predicted potential annual water yield reduction (mm/yr) due to the conversion of grasslands to forest lands, showing a strong increasing gradient from the dry and cold northwest to the warm and wet southeast. **(See color insert after p. 162.)** Regions with annual precipitation of less than 400 mm per year are not appropriate for reforestation and were excluded from the analysis. (From Sun et al. 2006, Potential water yield reduction due to reforestation across China. *Journal of Hydrology,* 328:548–558.)

This analysis was based on the assumption that future precipitation and potential evapotranspiration do not change. A changing climate will certainly result in a different scenario on forestation impacts. There is some evidence that overall ecosystem productivity has been increasing across China in the past decade (Fang et al. 2003). The increasing trend of productivity may indicate an increasing trend of water use because water is tightly coupled to ecosystem productivity in general (Jackson et al. 2005).

7.4 IMPLICATIONS OF FOREST–WATER RELATIONS TO FORESTATION CAMPAIGNS IN CHINA

Worldwide research on forest–water relations in the past few decades provides a basis for projecting the hydrologic consequences of forestation efforts. We now know that in general, forests provide the best water quality since soil erosion in undisturbed forests is extremely low. However, they do use more water than other nonirrigated crops that have less root mass and shallower rooting depth. Potential streamflow reduction from reforestation is of great concern (Jackson et al. 2005, Sun et al. 2006). Forestation activities have limited effects on volume and peaks of large floods. Also, there is much variability of hydrologic responses to forestation.

Based on reviewed literature, we expect large spatial and temporal variability of hydrologic response to forestation because of the large gradients in climate (Sun et al. 2006), topography, soils, degree of disturbances, and stage of vegetation recovery in China. Those factors are well discussed in Andreassian (2004) and Scott et al. (2005).

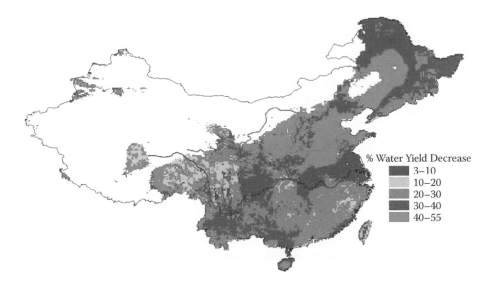

FIGURE 7.6 The potential water yield reduction as a percentage of water yield from previous grasslands following reforestation shows a strong decreasing gradient from the dry and cold northwest to the warm and wet southeast. **(See color insert after p. 162.)** Reforestation activities in the Yellow River basins will have a more pronounced impact than in the Yangtze River basins. Regions with annual precipitation of less than 400 mm per year are not appropriate for reforestation and were excluded from the analysis. (From Sun et al. 2006, Potential water yield reduction due to reforestation across China. *Journal of Hydrology*, 328:548–558.)

Although caution is needed to extrapolate studies from one region to others, existing literature has important implications for the current reforestation efforts in China.

1. *Forestation or converting from rain-fed croplands to tree plantations will likely reduce total annual streamflow.* Most literature clearly shows this conclusion because trees generally use more water than crops that have a short growing season and shallow rooting depth (Andreassian 2004). In China, exotic, fast-growing tree species such as larch, eucalyptus, and poplars are often used for timber production. Trees used for soil erosion control also often have economic considerations either for wood or fruit production. Those trees usually use more water than the native tree or shrub species.
2. Forestation is not likely to reduce stormflow volumes and peak flow rates, therefore forestation is not likely reduce large-scale floods. It is noteworthy to point out that a majority of the lands considered as viable candidates for forestation in China have chronic severe soil erosion problems. Such soils normally have degraded hydrologic properties that retard infiltration and therefore increase overland flow (Scott et al. 2005). Revegetation can improve soil properties such as increasing hydraulic conductivity and macroporosity. However, it may even take a long time for vegetation to affect soil infiltration capacity, and eventually stormflow peaks and volumes. Stormflow volumes and peak flow rates are mostly controlled by soil water storage capacity (i.e.,

soil depth and porosity). Large floods normally occur when the canopy and litter interception capacity and soil water storage have been filled, thus vegetation has very limited influences on flooding during large storm events. Antecedent soil moisture conditions are important when evaluating the role of forests in reducing peak flow rates. Brown et al. (2005) studied the seasonal effects of deforestation and forestation and concluded that those activities had higher impacts on baseflows than other types of flows.

3. Forestation is not likely to augment baseflows and spring occurrences. Baseflows are streamflows during non-rainfall periods originating from groundwater and soil water storage reservoirs. Compared to heavily degraded watersheds, it is generally true that undisturbed forests that have a thick litter layer and porous soils can store a greater amount of precipitation and release it gradually as spring waters. However, forestation on degraded lands is not likely to increase groundwater storage capacity and soil water storage in the short term. The increased filtration due to vegetation establishment may be exceeded by the increased water loss by evapotranspiration of the newly established forest (Scott et al. 2005). We predict that baseflow reductions are most pronounced in northern China where water stress is common throughout the growing season. Tree plantings on old floodplains and dried channel beds, and the Loess Plateau regions with deep soils are most likely to have impacts on groundwater recharge, soil moisture, and baseflow. Forestation in wetland-dominated watersheds may have little effect on overall watershed hydrology since water balances (i.e., evapotranspiration) are not likely to change significantly (Sun et al. 2000). Actual evapotranspiration in wetlands is generally close to potential evapotranspiration regardless of vegetation conditions. Therefore, in contradiction to the general perception that forests augment low flow or have more natural springs, forestation may actually reduce baseflow in the short term.

4. *The hydrologic effects of forestation will be small in the short term.* Deforestation has immediate effects on streamflow, but it takes longer for the trees to grow back to a mature forest (Brown et al. 2005). It takes even longer for degraded lands to develop into well-functioning forests across the temperate and boreal regions in northern China. It may take less time for tree establishment in warm, humid southern China where the climate is optimum for tree growth. However, nutrients often limit tree growth due to past chronic soil erosion. Therefore, we expect that the watershed hydrology of many newly forested sites will not cause large changes in the short term unless significant mechanical site preparation activities (i.e., terracing) have altered the soil hydrologic properties. This is especially true for degraded soils that have been chronically eroded and whose soil physical properties are damaged. A recent review on the impacts of mechanical disturbance on soil properties suggests that soil natural recovery from compaction may take several decades (National Council for Air and Stream Improvement [NCASI] 2004). A simple conceptual model was developed to illustrate the effects of forestation on water yield over time across major regions in China (Figure 7.7). The model suggests hydrologic recovery rates depend on climate, soil, and vegetation reestablishment.

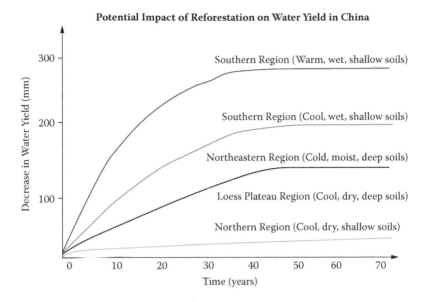

FIGURE 7.7 A conceptual model illustrating the gradual reductions of annual water yield following forestation across the major geographic regions in China. (From Sun et al. 2006, Potential water yield reduction due to reforestation across China. *Journal of Hydrology,* 328:548–558.)

Our discussion on forestation has been focused on the impact potential for basins subjected to complete cover change from bare lands or grasses and crops. This type of change is very unlikely to happen for large basins in China even under the current massive forestation campaigns because large areas of cropland are needed to meet the food demands in the rural areas. As shown in Figure 4 in Andreassian's (2004) review, the forestation effect on streamflow closely correlates to the percentage of land cover change. Our discussion has ignored the effects of soil and water conservation practices such as contouring, terracing and other bioengineering methods. Those practices may enhance infiltration, increase surface roughness, and consequently may have impacts on water balances, stormflow and baseflow characteristics.

7.5 FOREST HYDROLOGY RESEARCH NEEDS IN CHINA

Most of the existing forest hydrologic studies in China focused on single processes at the field scale (Zhang et al. 2000, Zhou et al. 2002, Zhou et al. 2004, Chen 1995). Integrated watershed-scale experiments and monitoring to evaluate the overall hydrologic response to forestation are rare. There is an urgent need to rigorously evaluate the hydrologic consequences of forestation, which is a key component of national "environmental reconstruction" (Zhou et al. 2001).

Andreassian (2004) raised seven important issues that require attention in studying forest–water relations at the watershed scale: (1) watershed size, (2) using models to mimic a control basin, (3) forest descriptors, (4) gradual changes, (5) long-term impacts, (6) distinguish forest stands from forest soils, and (7) number of watersheds. Those recommendations are quite pertinent to understanding and predicting the effects of reforestation in China.

Worldwide forest hydrology studies in the past century demonstrated that the paired watershed approach is the best way to detect land cover change effects on hydrology (Brown et al. 2005). The paired watershed approach removes the climatic variability effects on watershed hydrology between two watersheds that have different vegetation covers. However, to date there are no long-term paired watershed experiments that could give substantial answers on the impact of forests on watershed hydrology for any of the regions in China. Existing national ecological monitoring networks, such as China Ecological Research Network (CERN) and China Forestry Ecological Network (CFEN), promise to provide useful results on forestation impacts on hydrology. A paired watershed approach with a long-term plan should be adopted across China.

Another way of examining land use change on hydrology is by simulation models. Computer modeling has been well accepted by the hydrologic community as an effective way to examine individual hydrologic processes and separate the roles of various factors (soil, climate, and plant growth status) (Sun et al. 1998, Deng and Li 2003, Yu et al. 2003). Computer simulation models may play a key role since most of the rivers in China are not gauged, especially in remote areas where hydrologic characteristics are unique. However, development, parameterization, calibration, and validation for simulation models require a large amount of field data, which are often expensive to obtain. Models must be built upon quality experimental data and model simulations require accurate input climatic drivers and parameters. To fully understand the role of land cover change on the water cycles, such as precipitation, mesoscale distributed computer models are needed to account for the feedbacks between land and climate. Such types of models require even more close integration of remotely sensed spatial databases and energy and water balances. Meso-scale models are becoming increasing operational at the regional scale (Chen et al. 2005).

When evaluating the hydrologic effects of forestation on degraded lands in China, it is important to recognize the different roles of vegetation and soils (Bruijnzeel 2004, Wei et al. 2005). Systematic and long-term research is needed to document the recovery process of soil hydrologic properties along with changes of forest water use during the entire life cycle of plantation trees. In addition, the hydrologic effects of the soil and water conservation measures (e.g., contouring and terracing) that are often employed in forestation on degraded lands need to be evaluated. It is necessary to separate the roles of vegetation from engineering in influencing watershed hydrology to maximize the ecologic benefits of forestation and forestation planning.

The diverse physiographic regions in China provide an excellent location to test hypotheses generated elsewhere around the world. Several unique watershed settings such as the Loess Plateau (i.e., dry, deep soils) and the upper reaches of the Yangtze River (i.e., wet, cold, steep slopes, shallow soils) may have unique responses to forestation. A process-based approach is needed to address the delicate differential responses to vegetation management among these different geophysical conditions (Wilcox 2002).

ACKNOWLEDGMENTS

This study was supported by the U.S. Department of Agriculture Forest Service Southern Global Change Program, in collaboration with the Southern China Botany Garden,

Chinese Academy of Science. Partial support is also provided by the China Key Basic Research Program (973 Program) Influence and Control of Forest Vegetation Covers on Agricultural Ecological Environments in Western China (2002CB111502).

REFERENCES

Andreassian, V. 2004. Waters and forests: From historical controversy to scientific debate. *Journal of Hydrology* 291:1–27.

Beschta, R. L., M. R. Pyles, A. E. Skaugset, and C. G. Surfleet. 2000. Peakflow responses to forest practices in the western Cascades of Oregon, USA. *Journal of Hydrology* 233:102–120.

Bosch, J. M., and J. D. Hewlett. 1982. A review of catchment experiments to determine the effect of vegetation changes on water yield and evapotranspiration. *Journal of Hydrology* 55:3–23.

Brown, A. E., L. Zhang, T. A. McMahon, A. W. Western, and R. A. Vertessy. 2005. A review of paired catchment studies for determining changes in water yield resulting from alterations in vegetation. *Journal of Hydrology* 310:28–61.

Bruijnzeel, L. A., 1996. Predicting the hydrologic effects of land cover transformation in humid tropics: The need for integrated research. In *Amazonian deforestation and climate*, ed. J. H. Gash, C. A. Nobre, J. M. Roberts, and R. L. Victoria. Chichester, UK: J. Wiley, 15–55.

Bruijnzeel, L. A., 2004. Hydrological functions of tropical forests: Not seeing the soils for the trees. *Agriculture Ecosystems and Environment* 104:185–228.

Burt, T., and W. T. Swank. 2002. Forest and floods? *Geography Review* 15(5):37–41.

Buttle, J. M., I. F. Creed, and J. W. Pomeroy. 2000. Advances in Canadian forest hydrology, 1995–1998. *Hydrological Processes* 14(9):1551–1578.

Buttle, J. M., I. F. Creed, and R. D. Moore. 2005. Advances in Canadian forest hydrology, 1999–2003. *Hydrological Processes* 19:169–200.

Chen, L. 1995. Infiltration rates of loess soils in a plantation for soil and water conservation [in Chinese]. *Journal of Beijing Forestry University* 17(3):51–55.

Chen, M., D. Pollard, and E. J. Barron. 2005. Hydrologic processes in China and their association with summer precipitation anomalies. *Journal of Hydrology* 301:14–28.

Cornish, P. M., and R. A. Vertessy. 2001. Forest age–induced changes in evapotranspiration and water yield in eucalypt forest. *Journal of Hydrology* 242:43–63.

Deng, H., and X. Li. 2003. Simulation of hydrologic response to land cover changes [in Chinese]. *Scientia Geographica Sinica* 58(1):1–5.

Fang, J., A. Chen, C. Peng, S. Zhao, and L. Ci. 2001. Changes in forest biomass carbon storage in China between 1949 and 1998. *Science* 292:2320–2322.

Fang, J., S. Piao, C. B. Field, Y. Pan, Q. Guo, L. Zhou, and C. Peng. 2003. Increasing new primary production in China from 1982 to 1999. *Frontiers Ecological Environment* 6:293–297.

Farley, E. G., J. B. Jobbagy, and Jackson. 2005. Effects of afforestation on water yield: A global synthesis with implications for policy. *Globe Change Biology* (11):1565–1576.

Federer, C. A., and D. Lash. 1978. *BROOK: A hydrologic simulation model for eastern forested*. Research Report 19. Durham: Water Resources Research Center, University of New Hampshire.

Ffolliott, P., and D. P. Guertin, eds. 1987. *Forest Hydrologic Resources in China: An Analytical Assessment. Proceedings of a Workshop, Harbin, China. August 18 –23*. Washington, DC: United States Man and the Biosphere Program.

Hewlett, J. D. 1982. Forests and floods in the light of recent investigation. In *Hydrological processes of forested areas*. National Research Council of Canada Publication no. 20548, Ottawa: NRCC, 543–559.

Hibbert, A. R. 1967. Forest treatment effects on water yield. In *Forest hydrology, proceedings of a National Science Foundation Advanced Science Seminar*, ed. W. E. Sopper and H. W. Lull. Oxford: Pergamon Press, 527–543.

Ice, G. G., and J. D. Stednick. 2004. *A century of forest and wildland watershed lessons.* Bethesda, MD: Society of American Foresters, 287.

Jackson, R. B., 2005. Trading water for carbon with biological carbon sequestration. *Science* 1944–1947.

Jones, J. A., and G. E. Grant. 1996. Long-term stormflow responses to clearcutting and roads in small and large basins, western Cascades, Oregon. *Water Resources Research* 32:959–974.

Lei, J. 2002. China's implementation of six key forestry programs. http://www.newscientist.com/article.ns?id=dn2291.

Li, W. 2001. A summary of perspective of forest vegetation impacts on water yield [in Chinese]. *Journal of Natural Resources* 16:398–405.

Liu, C., and Y. Zeng. 2002. Effects of forests on water yield [in Chinese]. Special issue of *Hydraulic Research in China* (October):112–117.

Liu, C.-M., and J. Zhong. 1978. Effects of forests on annual streamflow in the Loess Plateau region [in Chinese]. *Acta Geographica Sinica* 33:112–126.

Liu, J., and J. Diamond. 2005. China's environment in a globalizing world. *Nature* 43:1179–1186.

Liu, J., H. Q. Tian, M. Liu, D. Zhuang, J. M. Melillo, and Z. Zhang. 2005. China's changing landscape during the 1990s: Large-scale land transformation estimated with satellite data. Geophysical Research Letters *32*: L02405, doi:10.1029/2004GL021649.

Ma, X. 1987. Hydrologic processes of a conifer forest in the sub-alpine region in Sichuan, China [in Chinese]. *Scientia Silvae Sinicae* 23:253–265.

Ma, X. 1993. *Forest hydrology* [in Chinese]. China Forest Publication House, 398.

NCASI (National Council for Air and Stream Improvement). 2004. *Effects of heavy equipment on physical properties of soils and long-term productivity: A review of literature and current research.* Technical Bulletin No. 887. Research Triangle Park, NC: National Council for Air and Stream Improvement, Inc.

Post, D. A,. and J. A. Jones. 2001. Hydrologic regimes of forested, mountainous, headwater basins in New Hampshire, North Carolina, Oregon, and Puerto Rico. *Advances in Water Resources* 24:1195–1210.

Robison, M. A.-L. et al. 2003. Studies of the impact of forests on peak flows and baseflows: A European perspective. *Forest Ecology and Management* 186:85–97.

Sahin, V., and M. J. Hall. 1996. The effects of afforestation and deforestation on water yields. *Journal of Hydrology* 178:293–309.

Scott, D. F., L. A. Bruijnzeel, and J. Mackensen. 2005. The hydrologic and soil impacts of reforestation. In *Forests, water and people in the humid tropics*, ed. M. Bonell and L. A. Bruijnzeel. Cambridge: Cambridge University Press, 622–651.

Scott, D. F., D. C. Le Maitre, and D. H. K. Fairbanks. 1998. Forestry and streamflow reductions in South Africa: A reference system for assessing extent and distribution. *Water SA* 24:187–199.

Stednick, J. D. 1996. Monitoring the effects of timber harvest on annual water yield. *Journal of Hydrology* 176:79–95.

Sun, G., S. G. McNulty, J. Lu, D. M. Amatya, Y. Liang, and R. K. Kolka. 2005. Regional annual water yield from forest lands and its response to potential deforestation across the southeastern United States. *Journal of Hydrology* 308:258–268.

Sun, G, S. G. McNulty, J. Moore, C. Bunch, and J. Ni. 2002. Potential impacts of climate change on rainfall erosivity and water availability in China in the next 100 years. Proceedings of the 12th International Soil Conservation Conference. Beijing, China, May 2002. 244–250.

Sun, G., H. Riekerk, and N. B. Comerford. 1998. Modeling the hydrologic impacts of forest harvesting on flatwoods. *Journal of American Water Resources Association* 34:843–854.

Sun, G., H. Riekerk, and L. V. Korhnak. 2000. Groundwater table rise after forest harvesting on cypress-pine flatwoods in Florida. *Wetlands* 20(1):101–112.

Sun, G., M. Riedel, R. Jackson, R. Kolka, D. Amatya, and J. Shepard. 2004. Influences of management of southern forests on water quantity and quality. In *Southern forest science: Past, present, and future*. Asheville, NC: U.S. Department of Agriculture, Forest Service, Southern Research Station, 195–224.

Sun, G., G. Zhou, Z. Zhang, X. Wei, S. G. McNulty, and J. M. Vose, 2006. Potential water yield reduction due to reforestation across China. *Journal of Hydrology* 328:548–558.

Swank, W. T., L. W. Swift Jr., and J. E. Douglass. 1988. Streamflow changes associated with forest cutting, species conversion, and natural disturbances. In *Ecological studies*. Vol. 66, *Forest hydrology and ecology at Coweeta*, ed. W. T. Swank and D. A. Crossley Jr. New York: Springer-Verlag, 297–312.

Thomas, R. B., and W. F. Megahan. 1998. Peak flow responses to clear-cutting and roads in small and large basins, western Cascades, Oregon: A second opinion. *Water Resources Research* 37(1):175–178.

Trimble, S. W., and F. H. Weirich. 1987. Reforestation reduces streamflow in the southeastern United States. *Journal of Soil and Water Conservation* 42(4): 274–276.

Vertessy, R. A., 1999. The impacts of forestry on streamflows: A review. In *Forest management for the protection of water quality and quantity: Proceedings of the second erosion in forests meeting, Warburton, Victoria, 4–6 May 1999*, ed. J. Croke and P. Lane. Cooperative Research Centre for Catchment Hydrology, Report 99/6, 93–109.

Vertessy, R. A. 2000. Impacts of plantation forestry on catchment runoff. In *Plantation, farm forestry and water: Proceedings of a national workshop, 20–21 July, Melbourne*, ed. E. K. Sadanandan Nabia and A. G. Brown, 9–19.

Vertessy, R. A., and Y. Bessard. 1999. Anticipation of the negative hydrologic effects of plantation expansion: Results from GIS-based analysis on the Murrumbidgee Basin. In *Forest management for the protection of water quality and quantity proceedings of the second Erosion in forests meeting, Warburton, Victoria, 4–6 May 1999*, ed. J. Croke, and P. Lane. 69–74.

Vertessy, R. A., F. Watson, and S. K. O'Sullivan. 2001. Factors determining relations between stand age and catchment water balance in mountain ash forests. *Forest ecology and management* 143:13–26.

Wang, L., and Z. Zhang. 1998. Advance in research on the eco-hydrologic effects of forests [in Chinese]. *World Forestry Research* 11(6):14–23.

Wei, X., W. Li, S. Liu, G.-Y. Zhou, and G. Sun. 2005b. Forest changes and streamflow—consistence and complexity [in Chinese]. *Journal of Natural Resources* 20(5):761–770.

Wei, X., S. Liu, G. Y. Zhou, and C. Wang. 2005a. Hydrological processes of key Chinese forests. *Hydrological Process* 19(1):63–75.

Wei, X., X. Zhou, and C. Wang. 2003. The influence of mountain temperate forest on the hydrology in northern China. *The Forestry Chronicle* 79:297–300.

Whitehead, P. G., and M. Robinson. 1993. Experimental basin studies—an international and historical perspective of forest impacts. *Journal of Hydrology* 145:217–230.

Wilcox, B.P. 2002. Shrub control and streamflow on rangelands: A process based viewpoint. *Journal of Range Management* 55:318–326.

Yang, H., L. Sun, and X. Yu. 1999. Water balances of three watersheds in the western Shanxi Province, China [in Chinese]. *Bulletin of Soil and Water Conservation* 14(2):26–31.

Yu, X. 1991. Forest hydrologic research in China. *Journal of Hydrology* 122:23–31.

Yu, X., G. Cheng, Y. Zhao, and Z. Zhang. 2003. Distributed hydrological model for forested watersheds [in Chinese]. *Science of Soil and Water Conservation* 1(1):35–40.

Zhang, P., G. Shao, G. Zhao, D. C. LeMaster, G. R. Parker, J. B. Dunning Jr., and Q. Li. 2000. China's forest policy for the 21st century. *Science* 288(5474):2135–2136.

Zhang, Z., L. X. Wang, X. Yu, and Y. Li. 2000. Effective roughness coefficients of overland flow for forested hill slopes [in Chinese]. *Scientia Silvae Sinicae* 36(5):22–27.

Zhang, Z., S. Wang, and L. Wang. 2004. Progress of forest hydrology research in China: A review. *Natural Resources* 16(1):71–78.

Zhou, G. Y., J. D. Morris, J. H. Yan, Z. Y. Yu, and S. L. Peng. 2002. Hydrological impacts of reforestation with eucalyptus and indigenous species: A case study in southern China. *Forest Ecology and Management* 167:209–222.

Zhou, G. Y., and X. Wei. 2002. Impacts of eucalyptus (*Eucalyptus exserta*) plantation on soil erosion in Guangdong Province, Southern China—a kinetic energy approach. *CATENA* 49(3):231–251.

Zhou, G. Y., G. Yin, J. Morris, J. Bai, S. Chen, G. Chu, and N. Zhang. 2004. Measured sap flow and estimated evapotranspiration of tropical *Eucalyptus urophylla* plantations in China. *Acta Botanica Sinica* 46:202–210.

Zhou, X, H. Zhao, and H. Sun. 2001. Proper assessment of the hydrologic effects of forests [in Chinese]. *Natural Resources* 16(5):420–426.

8 Application of TOPMODEL for Streamflow Simulation and Baseflow Separation

Pei Wen, Xi Chen, and Yongqin Chen

8.1 INTRODUCTION

The TOPMODEL concept (Beven and Kirkby 1979, Beven and Wood 1983, O'Loughlin 1986, Ambroise et al. 1996) has been a popular watershed modeling tool (e.g., Anderson et al. 1997, Lamb et al. 1998, Guntner et al. 1999, Scanlon et al. 2000). It is widely used because of its conceptual simplicity of runoff generation, innovative use of topographical data, and demonstrated applicability to a wide variety of situations. In recent years, however, various hydrologists have noted the inappropriateness of TOPMODEL's conceptual basis to meaningfully describe hydrologically shallow, hilly situations where transient, perched groundwater flow plays a substantial role in runoff generation processes (Moore and Thompson 1996, Woods et al. 1997, Frankenberger et al. 1999, Scanlon et al. 2000). Their observation in forested catchments has suggested the presence of such a storm flow zone perched above low-conductivity layers in the soil or a slowly moving wetting front (Hammermeister et al. 1982a). A transient occurrence of storm flow through the macroporous region of the shallow subsurface may result in the rapid rise of the hydrograph. Data collected from subsurface weirs (Scanlon et al. 2000) showed that this flow occurs quickly enough to contribute to peak stream discharge, and that a greater percentage of precipitation is converted to subsurface flow in the lower hill slopes.

Many efforts have been made to improve TOPMODEL structure in order to account for this runoff generation mechanism. Scanlon et al. (2000) believe that the soil water component may arise from saturated flow disconnected from the permanent water table while previous models relied on a conceptual model of one continuous water table (Robson et al. 1992, Hornberger et al. 1985), where stream water concentrations are determined by the position of this water table relative to an upper and a lower soil zone. They explicitly modified TOPMODEL to incorporate shallow, lateral subsurface flow by using two simultaneous TOPMODEL simulations, one describing deep baseflow and the other describing shallow interflow.

The main objective of this study is to simulate streamflow and to estimate baseflow in a hilly forest catchment in southeastern China using the modified TOPMODEL. The model is calibrated and validated on the basis of daily and hourly observed streamflow data. Simulated hydrological components reveal how much baseflow contributes to the total runoff in the study region.

8.2 MODIFIED TOPMODEL

The theory underlying the modified TOPMODEL relates hydrological behavior to the topography-derived variable $ln(a/tan\beta)$, where a is the area drained per unit contour, β is the local slope angle, and $ln(\ )$ is the Naperian logarithm. The model calculations are semidistributed in the sense that they are carried out for increments of $ln(a/tan\beta)$ for the catchment (Hornberger et al. 1985). TOPMODEL was modified to account for cases in which separate subsurface storm flow and groundwater storage mechanisms contribute to stream discharge, a generalized presentation of which is given by Clapp et al. (1996). The primary modification was the addition of a second subsurface state variable, although modifications to associated fluxes were consequently necessary. Vertical recharge to the groundwater zone is taken into account, and both the subsurface storm flow zone and the groundwater zone can contribute to episodic surface saturation near the stream (Figure 8.1).

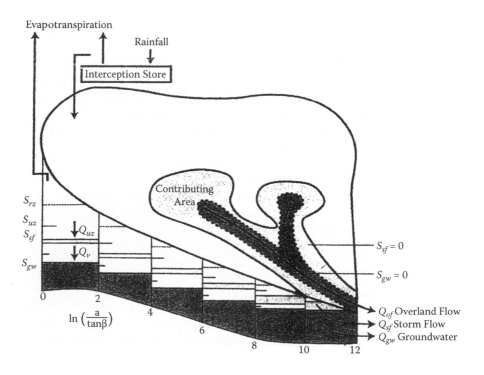

FIGURE 8.1   Schematic diagram of the modified TOPMODEL. (After Scanlon et al. 2000. Shallow subsurface storm flow in a forested headwater catchment: Observations and modeling using a modified TOPMODEL. *Water Resources Research* 36(9):2575–2586.)

8.2.1 Subsurface Flow

Following Beven and Wood (1983), the local groundwater saturated storage deficit S_{gwi} for any value of $ln(a/tan\beta)$ is related to the average catchment storage deficit, $\overline{S}_{gw}$, by

$$S_{gwi} = \overline{S}_{gw} + m[\Lambda - \ln(a/\tan\beta)_{gwi}] \tag{8.1}$$

where m is a scaling parameter, and Λ is the areal average of $ln(a/tan\beta)$.

The subsurface storm flow saturation deficit S_{sfi} is determined (Ambroise et al. 1996) by

$$S_{sfi} = S\max_{sf} - \frac{(a/\tan\beta)}{1/A\int_{A}(a/\tan\beta dA)}\left(S\max_{sf} - \overline{S}_{sf}\right) \tag{8.2}$$

where $S\max_{sf}$ is the maximum subsurface storm flow zone deficit, and $\overline{S}_{sf}$ is the average subsurface storm flow zone deficit for the catchment.

Surface saturation is controlled by the interaction of both subsurface deficits, $\overline{S}_{sf}$ and $\overline{S}_{gw}$ (Scanlon et al. 2000). Values of $S_{gwi} \leq 0$ and $S_{sfi} \leq 0$ indicate the area of groundwater saturation and storm flow zone saturation, respectively. For $S_{gw} \geq 0$ or $S_{sf} \geq 0$, the soil is partially unsaturated. Unsaturated zone calculations are made for each $ln(a/tan\beta)$ increment. The calculations use two storage elements, SUZ and SRZ. SRZ represents a root zone storage, the deficit of which is 0 at *field capacity* and becomes more positive as the soil dries out; SUZ denotes an unsaturated zone storage that is 0 at *field capacity* and becomes more positive as storage increases. Storage subject to drainage is represented by SUZ_i for the i-th increment of $ln(a/tan\beta)$. When $SUZ_i > 0$, vertical flow to the storm flow zone is calculated as

$$QUZ = \frac{SUZ_i}{S_{gwi}t_d} \tag{8.3}$$

where SUZ_i is the local unsaturated zone storage due to gravity drainage, and parameter t_d is a time constant.

Vertical drainage that depletes the water in the subsurface storm flow zone and replenishes the water stored in the groundwater zone (Scanlon et al. 2000) is expressed as

$$Q_v = \sum_{i=1}^{N} \min\left[c\left(S\max_{sf} - S_{sfi}\right), S_{gwi}\right]A_i \tag{8.4}$$

where $c[T^{-1}]$ is a simple transfer coefficient, N is the number of topographic index bins, and $A_i [L^2]$ is the fractional catchment area corresponding to each bin.

Evapotranspiration is taken from the SRZ_i store. The maximum value of storage in this zone is described as parameter $SRMAX$. The rate of evapotranspiration loss E is assumed to be proportional to a specified potential rate Ep and the root zone storage SRZ, as

$$E = Ep * SRZ / SRMAX \qquad (8.5)$$

The sum of vertical flows Q_v weighted by the area associated with each $ln(a/tan\beta)$ increment is added to reduce the average saturated deficit $\overline{S}_{gw}$. An outflow from the saturated groundwater zone, QB, is calculated as

$$QB = e^{-\Lambda} e^{\frac{\overline{S}_{gwi}}{m}} \qquad (8.6)$$

or

$$QB = Q_0 e^{-\frac{\overline{S}_{gwi}}{m}} \qquad (8.7)$$

where Q_0 is the initial stream discharge.

The average subsurface storm flow zone deficit $\overline{S}_{sf}$ changes over each simulation time step with inputs from overlying unsaturated zone Q_{uz}, and outflow to the stream Q_{sf}, and vertical drainage to the groundwater zone Q_v. Discharge from this zone is expressed as

$$Q_{sf} = Q_{0sf}\left(1 - \frac{\overline{S}_{sf}}{S\max_{sf}}\right) \qquad (8.8)$$

where Q_{0sf} is a storm flow zone recession parameter and influences the storm flow recession slope.

The water balance calculation for $\overline{S}_{gw}$ or $\overline{S}_{sf}$ produces a new end-of-time step value that is used to calculate a new value of S_{gwi} or S_{sfi} at the start of the next time step. There should be no water balance error involved since the incremental change in $\overline{S}_{gw}$ or $\overline{S}_{sf}$ is equal to the areally weighted sum of changes in S_{gwi} or S_{sfi}.

8.2.2 SURFACE FLOW

Surface flow may be generated either due to a calculated value of $S_{gwi} = 0$ or $S_{sfi} = 0$ in the saturated zone or due to the unsaturated zone deficit being satisfied by input from above ($SRZ = SRMAX$, $SUZ > S_{gwi}$ or $SUZ > S_{sfi}$ for any increment). Both cases represent saturation excess mechanisms of runoff production. Areas of high values of $ln(a/tan\beta)$, that is, areas of convergence or low slope angle, will saturate first and as the catchment becomes wetter, the area contributing surface flow will increase. Calculated surface flow at any time step is simply the water in excess of any deficit in each $ln(a/tan\beta)$ increment.

8.2.3 CHANNEL ROUTING

After both surface and subsurface water flows into the stream channel, it is routed through the channel system to the stream outlet. Routing should determine the number of time steps controlled by a specified maximum channel flow distance, D_{max}, and a constant channel wave velocity parameter, V. It is assumed that all runoff produced at each time step reaches the catchment outlet within a single time step.

8.3 APPLICATION

8.3.1 STUDY SITE

The study site is the Xingfeng catchment, which is situated in the forested headwater areas of the Dongjiang basin in South China. The catchment area is 42.6 km² and approximately 90% of the land surface is covered by forest. Soil is primarily red loam consisting of sandy loam and sand silt. Ground surface elevation varies from 42.6 to 508 m. The simulation program written by Beven and Wood (1983) is used for calculation of the topographic index on the basis of a DEM with a resolution of 25 m. Precipitation from five observation stations in Figure 8.2 is used to calculate the area's mean precipitation between 1982 and 1987. Additional data of pan evaporation and stream discharge from the observation station of the catchment outlet are used for model parameter calibration and model validation.

FIGURE 8.2 Map of the study catchment.

TABLE 8.1
Model parameters after calibration.

Parameter	Description	Value
m	Exponential storage parameter	0.16 m
SRMAX	Root zone available water capacity	0.22 m
td	Unsaturated zone time delay per unit storage deficit	0.35 h
$Smax_{sf}$	Maximum subsurface storm flow zone deficit	0.125 m
C	Recharge transfer coefficient	$1.15 \text{ m}^{-2}\text{h}^{-1}$
V	Catchment routing velocity	1650 m/h
T0	Mean catchment value of ln(T0)	$1.0 \text{ m}^2/\text{h}$
BC	Evaporation rate from root system	0.8
Alpha	Forestation coefficient	0.7
Silmax	Maximum water intercepted by leaf and litter cover	0.007 m

8.3.2 MODEL CALIBRATION AND VALIDATION

The observed daily precipitation, pan evaporation, and stream discharge from 1982 to 1985 are selected for model parameter calibration, and the daily data from 1986 to 1987 for model validation. Additionally, nine hourly flood events are chosen for the model simulation. The calibrated parameters are listed in Table 8.1. The Nash-Sutcliff efficiency coefficient (NSEC) is 0.79 and 0.72 in the calibration and validation periods, respectively. The root mean square error (RMSE) is 1.50 mm/d in the calibration period and 1.31 mm/d in the validation period. For the hourly simulation, five flood events are selected for model calibration and the other four for model validation. Calibration results demonstrate that most of the model parameters in hourly simulation are the same as those in daily simulations, except for the routing velocity (V), which becomes larger in flooding periods. V is 1,650 and 4,000 m/h in daily and hourly simulation, respectively. NSEC for all flood events is between 0.77 and 0.96 in the calibration period and between 0.81 and 0.87 in the validation period (Table 8.2). Figures 8.3 and 8.4 show that the simulated and observed streamflow discharges in the daily and hourly processes generally march well.

Based on the modified TOPMODEL structure, baseflow that comes from permanent groundwater storage can be estimated by equation (8.6) or (8.7). Estimation of mean baseflow from daily simulation is approximately 72% of the total discharge. Figure 8.5 shows the simulated hydrological components of surface flow, storm flow, and baseflow for the flood (No. 053008) during May 30 and 31, 1984. The subsurface stormflow discharge is about 15.6% of the total discharge and the baseflow is 23.4% in the flood event. For the nine flood events, calculated results in Table 8.2 demonstrate that the mean baseflow is 47.0% of the total discharge, and storm flow and surface flow are 7% and 46%, respectively.

TABLE 8.2
Simulation results of hourly flood discharges.

Flood number	Period	NSEC	RMSE (m³/s)	Flood discharge error (%)	Stormflow (%)	Baseflow (%)
052808	May 25–28, 1982	0.82	1.14	–2.9	7.0	50.2
020108	Feb. 1–3, 1983	0.86	1.46	7.3	4.0	52.6
052000	May 20–21, 1983	0.96	4.88	10.6	5.0	47.7
053008	May 30–31, 1984	0.77	4.58	–0.2	15.6	23.4
062500	Jun. 25–28, 1985	0.81	1.83	–1.9	1.0	42.2
070219	Jul. 2–4, 1985	0.90	3.31	5.4	6.0	38.9
091113	Sept. 11–13, 1985	0.87	2.35	9.1	1.0	49.4
092320	Sept. 23–26, 1985	0.88	2.58	–4.2	4.0	50.3
050712	May 7–24, 1987	0.81	15.84	12.1	16.0	68.7
Mean	Nine flood events	0.85	4.22	3.9	6.6	47.0

8.4 CONCLUSIONS

Traditionally, a main objective of hydrograph analysis is to decompose streamflow into the three major components of surface runoff, interflow, and baseflow. Quantifying time-dependent volumetric contributions to total stream water from surface and subsurface hydrological pathways is critical for the development of remedial strategies in areas where contaminants may be present and is necessary for the proper conceptualization of solute transport at the catchment scale. Over the past decade, chemical and isotopic methods for the separation of stream hydrographs have been a rigorously explored topic in the field of hydrology. Hydrological models used for this purpose must take into account and be consistent with site-specific observations of runoff-generating processes, and must be compatible with the theory derived from physical and geochemical observations in catchment studies. In this study, the modified TOPMODEL with an improvement in representing the runoff generation mechanism in the forest headwater area has successfully been applied in daily and hourly streamflow simulation in the Xingfeng catchment. Model calibration and validation results demonstrate that this model is able to effectively reflect watershed hydrological processes. By isolating the long-term groundwater recession and using a hydrograph transformation consistent with the TOPMODEL assumptions, the individual characteristics of both the subsurface storm flow and groundwater contributions to discharge have been evaluated. Simulation results demonstrate that baseflow is approximately 72% of the total discharge in the area, and therefore baseflow is very important for water resources utilization and further study on maintaining a basic baseflow is very important for environmental and ecosystem protection.

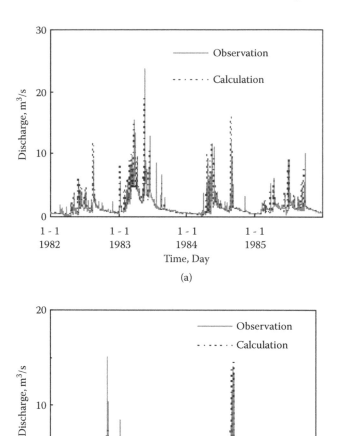

FIGURE 8.3 Observed and simulated daily discharges.

ACKNOWLEDGMENTS

The work described in this paper was supported by a grant from the Research Grants Council of the Hong Kong Special Administrative Region, China (Project No. CUHK4247/03H), and partially supported by open funding from the Key Lab of Poyang Lake Ecological Environment and Resource Development, Jiangxi Normal University and by the Program for New Century Excellent Talents in University, China (NCET-04-0492).

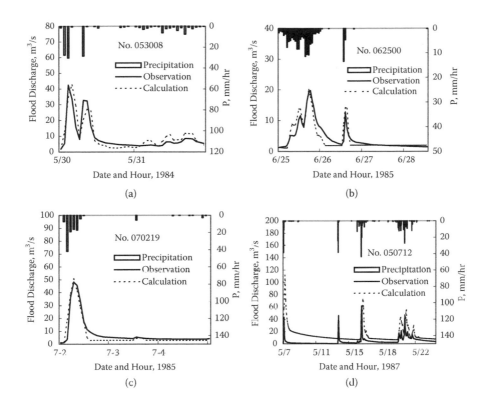

FIGURE 8.4   Observed and simulated hourly flood discharges.

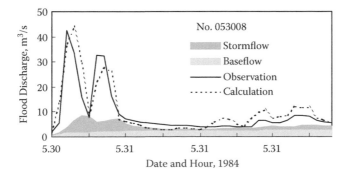

FIGURE 8.5   Simulated hydrological components of surface flow, stormflow, and baseflow.

REFERENCES

Ambroise, B., K. Beven, and J. Freer. 1996. Toward a generalization of the TOPMODEL concepts: Topographic indices of hydrological similarity. *Water Resources Research* 32(7):2135–2145.

Anderson, M., N. E. Peters, and D. Walling, eds. 1997. Special Issue: TOPMODEL, *Hydrological Processes* 11(9):1069–1356.

Beven, K. J., and M. J. Kirkby. 1979. A physically based variable contributing area model of basin hydrology. *Hydrological Sciences Bulletin* 24(1):43–69.

Beven, K. J., and E. Wood. 1983. Catchment geomorphology and the dynamics of runoff contributing areas. *Journal of Hydrology* 65:139–158.

Clapp, R. B., T. M. Scanlon, and S. P. Timmons. 1996. Modifying TOPMODEL to simulate the separate processes that generate interflow and baseflow. *Eos Trans. AGU*, 77(17), Spring Meet. Suppl., S122.

Frankenberger, J. R., E. S. Brooks, M. T. Walter, M. F. Walter, and T. S. Steenhuis. 1999. A GIS-based variable source area model. *Hydrological Processes* 13(6):804–822.

Güntner, A., S. Uhlembrook, J. Seibert, and Ch. Leibundgut. 1999. Multicriterial validation of TOPMODEL in a mountainous catchment. *Hydrological Processes* 13:1603–1620.

Hammermeister, D. P., G. F. Kling, and J. A. Vomocil. 1982a. Perched water tables on hillsides in western Oregon, I: Some factors affecting their development and longevity. *Soil Sci. Soc. Am. J.*, 46(4):811–818.

Hornberger, G. M., K. J. Beven, B. J. Cosby, and D. E. Sappington. 1985. Shenandoah watershed study: Calibration of a topography-based, variable contributing area hydrological model to a small forested catchment. *Water Resources Research* 21(12):1841–1850.

Lamb, R., K. J. Beven, and S. Myrabø. 1998. Use of spatially distributed water table observations to constrain uncertainty in a rainfall-runoff model. *Advances in Water Resources* 22(4):305–317.

Moore, R. D., and J. C. Thompson. 1996. Are water table variations in a shallow forest soil consistent with the TOPMODEL concept? *Water Resources Research* 32(3):663–669.

O'Loughlin, E. M. 1986. Prediction of surface saturation zones in natural catchments by topographic analysis. *Water Resources Research* 22:794–804.

Robson, A., K. J. Beven, and C. Neal. 1992. Towards identifying sources of subsurface flow: A comparison of components identified by a physically based runoff model and those determined by chemical mixing techniques. *Hydrol. Processes* 6:199–214.

Scanlon, T. M., J. P. Raffensperger, G. M. Hornberger, and R. B. Clapp. 2000. Shallow subsurface storm flow in a forested headwater catchment: Observations and modeling using a modified TOPMODEL. *Water Resources Research* 36(9):2575–2586.

Walter, M. T, T. S. Steenhuis, V. K. Mehta, D. Thongs, M. Zion, and E. Schneiderman. 2002. Refined conceptualization of TOPMODEL for shallow subsurface flows. *Hydrol. Process.* 16: 2041–2046.

Woods, R. A., M. Sivapalan, and J. S. Robinson. 1997. Modeling the spatial variability of subsurface runoff using a topographic index. *Water Resources Research* 33(5):1061–1073.

9 Spatially Distributed Watershed Model of Water and Materials Runoff

Thomas E. Croley II and Chansheng He

9.1 INTRODUCTION

Agricultural nonpoint source contamination of water resources by pesticides, fertilizers, animal wastes, and soil erosion is a major problem in much of the Laurentian Great Lakes Basin, located between the United States and Canada. Point source contaminations, such as combined sewerage overflows (CSOs), also add wastes to water flows. Soil erosion and sedimentation reduce soil fertility and agricultural productivity, decrease the service life of reservoirs and lakes, and increase flooding and costs for dredging harbors and treating wastewater. Improper management of fertilizers, pesticides, and animal and human wastes can cause increased levels of nitrogen, phosphorus, and toxic substances in both surface water and groundwater. Sediment, waste, pesticide, and nutrient loadings to surface and subsurface waters can result in oxygen depletion and eutrophication in receiving lakes, as well as secondary impacts such as harmful algal blooms and beach closings due to viral and bacterial and/or toxin delivery to affected sites. The U.S. Environmental Protection Agency (EPA) has identified contaminated sediments, urban runoff and storm sewers, and agriculture as the primary sources of pollutants causing impairment of Great Lakes shoreline waters (USEPA 2002). Prediction of various ecological system variables or consequences (such as beach closings), as well as effective management of pollution at the watershed scale, require estimation of both point and nonpoint source material transport through a watershed by hydrological processes. However, currently there are no integrated fine-resolution spatially distributed, physically based watershed-scale hydrological/water quality models available to evaluate movement of materials (sediments, animal and human wastes, agricultural chemicals, nutrients, etc.) in both surface and subsurface waters in the Great Lakes watersheds.

The Great Lakes Environmental Research Laboratory (GLERL) and Western Michigan University are developing an integrated, spatially distributed, physically-based hydrology and water quality model. It is a nonpoint source runoff and water quality model used to evaluate both agricultural nonpoint source loading from soil erosion, fertilizers, animal manure, and pesticides, and point source loadings at the watershed level. GLERL is augmenting an existing physically based distributed surface/subsurface hydrology model (their Distributed Large Basin Runoff Model) by adding material transport capabilities to it. This will facilitate effective Great Lakes watershed management decision making, by allowing identification of critical risk areas and tracking of different sources of pollutants for implementation of water quality programs, and will augment ecological prediction efforts. This paper briefly reviews distributed watershed models of water and agricultural materials runoff and identifies their limitations, and then presents our resultant distributed model of water and material movement within a watershed.

9.2 AGRICULTURAL RUNOFF MODELS

Estimating point and nonpoint source pollutions and CSOs is critical for planning and enforcement agencies in protection of surface water and groundwater quality. During the past four decades, a number of simulation models have been developed to aid in the understanding and management of surface runoff, sediment, nutrient leaching, and pollutant transport processes. The widely used water quality models include ANSWERS (Areal Nonpoint Source Watershed Environment Simulation) (Beasley and Huggins 1980), CREAMS (Chemicals, Runoff, and Erosion from Agricultural Management Systems) (Knisel 1980), GLEAMS (Groundwater Loading Effects of Agricultural Management Systems) (Leonard et al. 1987), AGNPS (Agricultural Nonpoint Source Pollution Model) (Young et al. 1989), EPIC (Erosion Productivity Impact Calculator) (Sharpley and Williams 1990), and SWAT (Soil and Water Assessment Tool) (Arnold et al. 1998) to name a few. These models all use the SCS Curve Number method, an empirical formula for predicting runoff from daily rainfall. Although the Curve Number method has been widely used worldwide, it is an event-based (storm hydrograph) method not really suitable for continuous simulations. Researchers have expressed concern that it does not reproduce measured runoff from specific storm rainfall events because the time distribution is not considered (Kawkins 1978; Wischmeier and Smith 1978; Beven 2000; Garen and Moore 2005). Limitations of the Curve Number method also include (1) no explicit account of the effect of the antecedent moisture conditions in runoff computation, (2) difficulties in separating storm runoff from the total discharge hydrograph, and (3) runoff processes not considered by the empirical formula (Beven 2000; Garen and Moore 2005). Consequently, estimates of runoff and infiltration derived from the Curve Number method may not well represent the actual. As sediment, nutrient, and pesticide loadings are directly related to infiltration and runoff, use of the Curve Number method may also result in incorrect estimates of nonpoint source pollution rates.

Due to the limitations of the Curve Number method, ANSWERS, CREAMS, GLEAMS, AGNPS, and SWAT were developed to assess impacts of different agricultural management practices, not to predict exact pesticide, nutrient, and sediment loading in a study area (Ghadiri and Rose 1992; Beven 2000; Garen and Moore 2005). In addition, most water quality models, such as CREAMS and GLEAMS, are field-size models and cannot be used directly at the watershed scale. Applications of these models have been limited to field-scale or small experimental watersheds. Some models, such as ANSWERS, CREAMS, EPIC, and AGNPS, also do not consider subsurface and groundwater processes.

Recently, several water quality models have been modified to take into consideration available multiple physical and agricultural databases. The USEPA designated two of the most widely used water quality models, SWAT and HSPF (Hydrologic Simulation Program in FORTRAN) (Bicknell et al. 1996), for simulation of hydrology and water quality nationwide. SWAT is a comprehensive watershed model and considers runoff production, percolation, evapotranspiration, snowmelt, channel and reservoir routing, lateral subsurface flow, groundwater flow, sediment yield, crop growth, nitrogen and phosphorous, and pesticides. But it uses the curve number method for estimating runoff, and therefore has the same limitations the curve number method has in runoff simulation. The basic simulation unit in SWAT is the sub-watershed, instead of a fine-resolution grid network, thus limiting its incorporation of spatial variability in simulating hydrologic processes.

Evolved from the Stanford Watershed Model (Crawford and Linsley 1966), HSPF is one of the most extensively used general hydrologic and water quality models (Bicknell et al. 1996). Under the auspices of the USEPA, the first version of the HSPF was completed in 1980. Since then, the model has gone through extensive revisions, corrections, refinements, and validations in many areas, and is one of the three simulation models included in BASINS (Better Assessment Science Integrating Point and Nonpoint Sources), the USEPA's watershed modeling tools for support of water quality management programs throughout the country (Lahlou et al. 1998). HSPF utilizes time series meteorology data to simulate hydrological processes in both pervious and impervious land segments. The hydrological processes in the model include accumulation and melting of snow and ice, water budget, sediment transport, soil moisture, and temperature. The water quality modules of the model include concentration and transport of nitrogen, phosphorus, pesticides, and other pollutants. However, HSPF requires extensive input parameters such as wind speed, dew point temperature, potential evapotranspiration, and channel characteristics. Many of these parameters are not available in most watersheds, particularly large watersheds. In addition, HSPF is a semidistributed model since a basin is divided into lumped-parameter model applications to subbasins and land parcels to coarsely represent spatial variations of rainfall and land surface. Moreover, neither SWAT nor HSPF considers nonpoint sources from animal manure and CSOs and infectious diseases. Thus, there is an urgent need for the development of a spatially distributed, physically based watershed model that simulates both point and nonpoint source pollutions in the Great Lakes Basin.

9.3 DISTRIBUTED LARGE BASIN RUNOFF MODEL

GLERL developed a large basin runoff model in the 1980s for estimating daily rainfall/runoff relationships on each of the 121 large watersheds surrounding the Laurentian Great Lakes (Croley 2002). It is physically based to provide good representations of hydrologic processes and to ensure that results are tractable and explainable. It is a lumped-parameter model of basin outflow consisting of a cascade of moisture storages or "tanks," each modeled as a linear reservoir, where tank outflows are proportional to tank storage. We applied it to a 1-km^2 "cell" of a watershed and modified it to allow lateral flows between adjacent cells for moisture storage; see

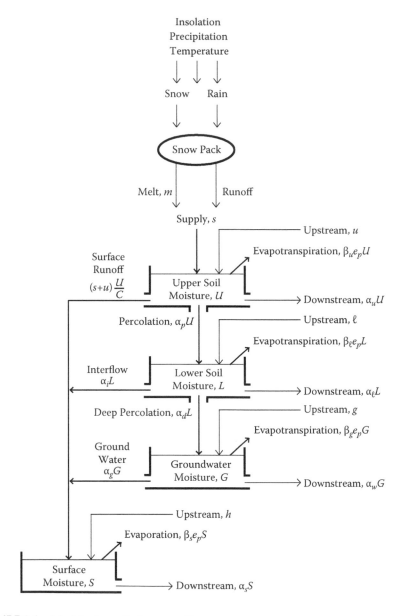

FIGURE 9.1   Model schematic for one cell.

Figure 9.1. By grouping cell applications appropriately, we built a spatially distributed accounting of moisture in several layers (zones), the distributed large basin runoff model (DLBRM). Daily precipitation, air temperature, and insolation (the latter available from cloud cover and meteorological summaries as a function of location and time of the year) may be used to determine snowpack accumulations, snowmelt (degree-day computations), and supply, s, into the upper soil zone. Water flow, u, also enters from upstream cells' upper soil zones. The total supply is divided into surface runoff, $(s+u)U/C$, and infiltration to the upper soil zone, $(s+u)(1-U/C)$, in relation to the upper soil zone moisture content, U, and the fraction it represents of the upper soil zone capacity, C (variable area infiltration). Percolation to the lower soil zone, $\alpha_p U$, evapotranspiration, $\beta_u e_p U$, and lateral flow to a downstream upper soil zone, $\alpha_u U$, are taken as outflows from a linear reservoir (flow is proportional to storage). Likewise, water flow, ℓ, enters the lower soil zone from upstream cells' lower soil zones. Interflow from the lower soil zone to the surface, $\alpha_i L$, evapotranspiration, $\beta_\ell e_p L$, deep percolation to the groundwater zone, $\alpha_d L$, and lateral flow to a downstream lower soil zone, $\alpha_\ell L$, are linearly proportional to the lower soil zone moisture content, L. Water flow, g, enters the groundwater zone from upstream cells' groundwater zones. Groundwater flow, $\alpha_g G$, evapotranspiration from the groundwater zone, $\beta_g e_p G$, and lateral flow to a downstream groundwater zone, $\alpha_w G$, are linearly proportional to the groundwater zone moisture content, G. Finally, water flow, h, enters the surface zone from upstream cells' surface zones. Evaporation from the surface storage, $\beta_s e_p S$, and lateral flow to a downstream surface zone, $\alpha_s S$, are linearly proportional to the surface zone moisture, S. Additionally, evaporation and evapotranspiration are dependent on potential evapotranspiration, e_p, as determined independently from a heat balance over the watershed, appropriate for small areas. The *alpha* coefficients (α) represent linear reservoir proportionality factors and the *beta* coefficients (β) represent partial linear reservoir coefficients associated with evapotranspiration. From Figure 9.1,

$$\frac{d}{dt}U = s + u - (s+u)\frac{U}{C} - \alpha_p U - \alpha_u U - \beta_u e_p U \tag{9.1}$$

$$\frac{d}{dt}L = \alpha_p U - \alpha_i L - \alpha_d L - \alpha_\ell L + \ell - \beta_\ell e_p L \tag{9.2}$$

$$\frac{d}{dt}G = \alpha_d L - \alpha_g G - \alpha_w G + g - \beta_g e_p G \tag{9.3}$$

$$\frac{d}{dt}S = (s+u)\frac{U}{C} + \alpha_i L + \alpha_g G - \alpha_s S + h - \beta_s e_p S \tag{9.4}$$

Solution

Consideration of equations (9.1)–(9.4) reveals multiple analytical solutions; while tractable, a simpler approach uses a numerical solution based on finite difference approximations of equations (9.1)–(9.4). Consider equation (9.1) approximated with finite differences,

$$\Delta U \cong (\bar{s}+\bar{u})\Delta t - \left(\frac{(\bar{s}+\bar{u})}{C} + \alpha_p + \alpha_u + \beta_u \bar{e}_p\right)\bar{U}\Delta t \tag{9.5}$$

where ΔU = change in upper soil zone moisture storage over time interval Δt, $\bar{s}$, $\bar{u}$, and $\bar{e}_p$ = average supply, upstream inflow, and potential evapotranspiration rates, respectively, over time interval Δt, and $\bar{U}$ = average upper soil zone moisture storage over time interval Δt. By taking $\Delta U = U - U_0$ (where U_0 and U are beginning-of- and end-of-time-interval storages, respectively) and $\bar{U} \cong U$, equation (9.5) becomes

$$U \cong \frac{U_0 + (\bar{s} + \bar{u})\Delta t}{1 + \left(\dfrac{\bar{s} + \bar{u}}{C} + \alpha_p + \alpha_u + \beta_u \bar{e}_p\right)\Delta t} \tag{9.6}$$

Equation (9.6) is good for small Δt and as $\Delta t \rightarrow 0$, equation (9.6) approaches the true solution (converges) to equation (9.1). Likewise, using similarly defined terms, equations (9.2)–(9.4) become

$$L \cong \frac{L_0 + (\alpha_p U + \bar{\ell})\Delta t}{1 + (\alpha_i + \alpha_d + \alpha_\ell + \beta_\ell \bar{e}_p)\Delta t} \tag{9.7}$$

$$G \cong \frac{G_0 + (\alpha_d L + \bar{g})\Delta t}{1 + (\alpha_g + \alpha_w + \beta_g \bar{e}_p)\Delta t} \tag{9.8}$$

$$S \cong \frac{S_0 + \left(\dfrac{\bar{s} + \bar{u}}{C}U + \alpha_i L + \alpha_g G + \bar{h}\right)\Delta t}{1 + (\alpha_s + \beta_s \bar{e}_p)\Delta t} \tag{9.9}$$

As equations (9.6)–(9.9) are used over time interval Δt, end-of-time-interval values are computed from beginning-of-time-interval values (e.g., U from U_0). These end-of-time-interval values for one time interval become beginning-of-time-interval values for the subsequent time interval.

Each cell's inflow hydrographs must be known before its outflow hydrograph can be modeled; therefore we arranged calculations in a flow network to assure this. It is determined automatically from a watershed map of cell flow directions. The flow network is implemented to minimize the number of pending hydrographs in computer storage and the time required for them to be in computer storage. We used the same network for surface, upper soil, lower soil, and groundwater storages. We implemented routing network computations as a recursive routine to compute outflow, which calls itself to compute inflows (which are upstream outflows) (Croley and He 2005, 2006).

9.3.1 Application

We have discretized 18 watersheds to date. The elevation map for the Kalamazoo River watershed in southwestern Michigan is shown in Figure 9.2. We used elevations taken from a 30-m digital elevation model (DEM) available from the United

Saugatuck,
Michigan, USA
86° 13′ W. Lon.
42° 40′ N. Lat.

612 km east-west
332 km north-south

N

180.00   360.00

FIGURE 9.2 Kalamazoo watershed elevations (m).

States Geological Survey (USGS). We also used USGS land cover characteristics
and the U.S. Department of Agriculture State Soil Geographic Database to add land
cover, upper and lower soil zone parameters (depth, actual water content, and permeability), soil texture, and surface roughness; see Croley et al. (2005). In application,
we used gradient search techniques to minimize root mean square error between
modeled and actual basin outflow by selecting the best spatial averages for each
of the eleven parameters; the spatial variation of each parameter follows a selected
watershed characteristic, as shown here and arrived at by experimentation.

$$\left(\alpha_p\right)_i = \overline{\alpha}_p \; f(K_i^U, 80\%) \tag{9.10}$$

$$\left(\beta_u\right)_i = \overline{\beta}_u \; f(K_i^U, 80\%) \tag{9.11}$$

$$\left(\alpha_i\right)_i = \overline{\alpha}_i \; f(K_i^L, 80\%) \tag{9.12}$$

$$\left(\alpha_d\right)_i = \overline{\alpha}_d \; f(K_i^L, 80\%) \tag{9.13}$$

$$\left(\beta_\ell\right)_i = \overline{\beta}_\ell \; f(K_i^L, 80\%) \tag{9.14}$$

$$\left(\alpha_g\right)_i = \overline{\alpha}_g \; f(K_i^L, 80\%) \tag{9.15}$$

$$\left(\alpha_s\right)_i = \overline{\alpha}_s \; f\!\left(\frac{\sqrt{s_i}}{\eta_i}, 80\%\right) \tag{9.16}$$

$$\left(\alpha_u\right)_i = \overline{\alpha}_u \; f(K_i^U, 80\%) \tag{9.17}$$

$$\left(\alpha_\ell\right)_i = \overline{\alpha}_\ell \; f(K_i^L, 80\%) \tag{9.18}$$

$$\left(\alpha_w\right)_i = \overline{\alpha}_w \; f(K_i^L, 80\%) \tag{9.19}$$

$$(C)_i = \bar{C} \, f(C_i^U, 80\%) \qquad (9.20)$$

$$f(x_i, \varepsilon) = \left(\frac{x_i}{\dfrac{1}{n} \displaystyle\sum_{j=1}^{n} x_j} - 1 \right) \frac{\varepsilon}{100\%} + 1 \qquad (9.21)$$

where $(\alpha_\bullet)_i$ = linear reservoir coefficient for cell i, $\bar{\alpha}_\bullet$ = spatial average value of the linear reservoir coefficient (from parameter calibration), $(\beta_\bullet)_i$ and $\bar{\beta}_\bullet$ are defined similarly for partial linear reservoir coefficients (used in evapotranspiration), $(C)_i$ and $\bar{C}$ are defined similarly for the upper soil zone capacity, K_i^U = upper and K_i^L = lower soil zone permeability in cell i, s_i = slope of cell i, η_i = Manning's roughness coefficient for cell i, C_i^U = upper soil zone available water capacity, x_i = data value for cell i, and n = number of cells in the watershed.

Note two parameters not shown here, which govern the heat balance used for snowmelt and potential evapotranspiration, are taken as spatially constant over the watershed. Also, the partial linear reservoir coefficients for the groundwater and surface zones are taken as zero, ignoring evapotranspiration from those two zones. Thus there are 13 parameters (of a possible 15) searched in the calibration. To speed up calibrations, we preprocessed all meteorology for all watershed cells and preloaded it into computer memory. The correlation between modeled and observed watershed outflows was 0.88, the root mean square error was 0.19 mm/d (compare with a mean flow of 0.78 mm/d); the ratio of modeled to actual mean flow was 1.00, and the ratio of modeled to actual flow standard deviation was 0.87 (Croley and He 2006). We used the model to look at modeling alternatives, including alternative evapotranspiration calculations, spatial parameter patterns, and solar insolation estimates. We also explored scaling effects in using lumped parameter model calibrations to calculate initial distributed model parameter values (Croley and He 2005; Croley et al. 2005).

9.3.2 Testing

As a test of equations (9.6)–(9.9), we used them for $\Delta t = 1.5$ minutes to approximate the solution of equations (9.1)–(9.4) over about 17 years of daily values for the Maumee River watershed (Croley and He 2006) and found them identical (in all variables) through three significant digits (all that were inspected) with the exact analytical solution. For $\Delta t = 15$ minutes, the solution was nearly identical with only an occasional difference of one in the third significant digit. As the Maumee River watershed has a very "flashy" response to precipitation (very fast upper soil and surface storage zones) these comparisons are deemed significant and the time intervals should be more than adequate for the slower response of lower soil and groundwater zones (the Maumee application has no lower soil or groundwater zones).

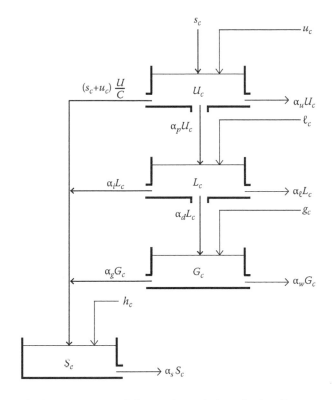

FIGURE 9.3   Distributed "pollutant" flows schematic for a single cell.

9.4 MATERIALS RUNOFF MODEL

Consider now the addition of some material or pollutant dissolved in, or carried by, the water flows in Figure 9.1, except that none is considered to be evaporated; see Figure 9.3. At any time, let the concentration of this conservative pollutant in the inflow u be c_u and in the supply s be c_s. If these flows do not mix together, then the fraction U/C of each of these flows runs off directly (without even entering the upper soil zone) and the surface runoff of pollutant is $(sc_s + uc_u)U/C$. If the concentration in the upper soil zone moisture storage U is c_U, then the percolating pollutant is $\alpha_p U c_U$ and the lateral pollutant flow downstream to the next cell's upper soil zone is $\alpha_u U c_U$. Taking pollutant movement with evaporation as zero, mass continuity (of the pollutant) gives:

$$\frac{d}{dt}(Uc_U) = sc_s + uc_u - (sc_s + uc_u)\frac{U}{C} - \alpha_p U c_U - \alpha_u U c_U \tag{9.22}$$

or

$$\frac{d}{dt}U_c = s_c + u_c - (s_c + u_c)\frac{U}{C} - \alpha_p U_c - \alpha_u U_c \tag{9.23}$$

where $s_c = sc_s$, $u_c = uc_u$, and $U_c = Uc_U$.

Likewise from Figure 9.3, mass continuity of the pollutant gives:

$$\frac{d}{dt}L_c = \alpha_p U_c - \alpha_i L_c - \alpha_d L_c - \alpha_\ell L_c + \ell_c \tag{9.24}$$

$$\frac{d}{dt}G_c = \alpha_d L_c - \alpha_g G_c - \alpha_w G_c + g_c \tag{9.25}$$

$$\frac{d}{dt}S_c = \left(s_c + u_c\right)\frac{U}{C} + \alpha_i L_c + \alpha_g G_c - \alpha_s S_c + h_c \tag{9.26}$$

where L_c, G_c, and S_c are the amounts of pollutant in the lower soil zone, the groundwater zone, and surface storage, respectively, and ℓ_c, g_c, and h_c are the upstream pollutant flows from the lower soil zone, the groundwater zone, and surface storage, respectively.

Solution

Similar to the numerical solution of equations (9.1)–(9.4) [(9.6)–(9.9)], the numerical solution for equations (9.23)–(9.26) becomes

$$U_c \cong \frac{U_{c0} + \left(\bar{s}_c + \bar{u}_c\right)\Delta t - \left(\bar{s}_c + \bar{u}_c\right)\dfrac{U}{C}\Delta t}{1 + \left(\alpha_p + \alpha_u\right)\Delta t} \tag{9.27}$$

$$L_c \cong \frac{L_{c0} + \left(\alpha_p U_c + \bar{\ell}_c\right)\Delta t}{1 + \left(\alpha_i + \alpha_d + \alpha_\ell\right)\Delta t} \tag{9.28}$$

$$G_c \cong \frac{G_{c0} + \left(\alpha_d L_c + \bar{g}_c\right)\Delta t}{1 + \left(\alpha_g + \alpha_w\right)\Delta t} \tag{9.29}$$

$$S_c \cong \frac{S_{c0} + \left(\dfrac{\bar{s}_c + \bar{u}_c}{C}U + \alpha_i L_c + \alpha_g G_c + \bar{h}_c\right)\Delta t}{1 + \alpha_s \Delta t} \tag{9.30}$$

where terms are defined for material flows in a manner similar to that for water flows. We used the same network for surface, upper soil, lower soil, and groundwater storage of pollutant as we used for water flows.

9.4.1 Initial and Boundary Conditions

Suppose a pollutant deposit P exists on top of the upper soil zone. Precipitation or snowmelt on top of this deposit will produce a supply s to the upper soil zone that will dissolve some of this pollutant, producing a pollutant concentration c_s. If we regard this process as independent of other flows to the top of the upper soil zone (u and $u_c = uc_u$), we can model the pollutant uptake as follows.

$$\frac{d}{dt}P = -sc_s, \quad \text{if } P > 0$$
$$= 0, \quad \text{if } P = 0$$
(9.31)

The finite difference solution is

$$P = P_0 - \bar{s}c_s\Delta t, \quad \text{if } \bar{s}c_s\Delta t < P_0$$
$$= 0, \quad \text{if } \bar{s}c_s\Delta t \geq P_0$$
(9.32)

where P and P_0 are the end- and beginning-of-time-interval pollutant values, respectively. The pollutant delivered to the top of the upper soil zone would be $\bar{s}_c + \bar{u}_c = \bar{s}c_s + \bar{u}c_u$ as used in equations (9.27) and (9.30).

9.5 EXAMPLE SIMULATION

Selected lateral flows simulated by the model for the upper soil zone, groundwater zone, and surface zone are shown in Figure 9.4 for the Kalamazoo River watershed for the first two months of 1952. Although the simulation was daily, the flows are shown weekly in Figure 9.4. The initial conditions for this simulation included 1 cm of (arbitrary) pollutant on the surface of the watershed on 1 January 1952. Within two weeks it was gone from the surface (@ $c_s = 1.0$ m^3 of pollutant / m^3 of water). Columns 1 through 3 in Figure 9.4 show that lateral water flows in the watershed were fairly uniform for the period with perhaps higher surface water flows on 1 and 15 January (more of the streamflow network is seen to respond then). Column 4 illustrates that even at the end of day 1, the upper soil zone already had pollutant in it; uper soil zone (USZ) pollutant flows peaked on 15 January and slowly tapered off through the end of February. Pollutant did not appear in any sizable manner in the groundwater zone until the end of February, as illustrated in column 5 of Figure 9.4. The pollutant flow in the surface network, shown in the last column in Figure 9.4, responds midway between the USZ and ground water zone (GWZ) responses. There is a high flush on the first day, corresponding with the initial condition (placing pollutant on the surface of the watershed at the beginning of day 1) and the increased surface water flow; the surface pollutant map shows extensive response. The surface response then drops off as pollutant is only available in the USZ and GWZ. However, on 15 January, the surface pollutant response increases with the flush of water through the USZ that occurs then (see third row in columns 3 and 4 in Figure 9.4).

9.6 SUMMARY

Prediction and management of watershed water quality require estimation of non-point source material movement throughout the watershed. We briefly reviewed

FIGURE 9.4 Kalamazoo model lateral flows (cm/d).

distributed agricultural runoff models and learned that there are no integrated spatially distributed, physically based watershed-scale hydrological/water quality models available to evaluate movement of materials in both surface and subsurface waters. Either the hydrology is limited to very simple empirical descriptions or the application is made to only very coarse spatial discretizations of the watershed. We adapted an existing lumped-parameter conceptual water balance model of watershed hydrology into a spatially distributed model of runoff. It employs moisture storage in the upper and lower soil zones, in a groundwater zone, and on the surface, with lateral flows from all storages into similar storages in adjacent grid cells defined over the watershed. By applying the surface drainage network to all storage lateral flows, we can trace the movement of water throughout the watershed. We further adapt the distributed model to incorporate the storage and movement of an arbitrary material, conservative in nature, and to trace its movement throughout the watershed. By employing a numerical solution instead of the analytical solution used in the original lumped-parameter water balance model, we are able to easily represent the mass balance of both water and an arbitrary conservative pollutant spatially throughout all storage zones in the watershed. Model testing reveals that the numerical solution converges to the analytical solution for a 1-km^2 grid on a watershed with a very fast response. By assigning initial pollutant surface amounts and introducing a single parameter (pollutant concentration in water), we can model its movement. In a simple example on the Kalamazoo River watershed, in which a uniform layer of pollutant is assumed initially, we present the consecutive spatial distributions that occur over a two-month simulation, demonstrating that the model could be used to simulate real-world material movement in a watershed.

ACKNOWLEDGMENTS

This is GLERL Contribution No. 1375.

REFERENCES

Arnold, G., R. Srinavasan, R. S. Muttiah, and J. R. Williams. 1998. Large area hydrologic modeling and assessment. Part I: Model development. *Journal of the American Water Resources Association* 34: 73–89.

Beasley, D. B., and L. F. Huggins. 1980. *ANSWERS (Areal Nonpoint Source Watershed Environment Simulation): User's manual*. West Lafayette, IN: Department of Agricultural Engineering, Purdue University.

Beven, K. J. 2000. *Rainfall-runoff modeling: The primer*. New York: John Wiley & Sons.

Bicknell, B. R., J. C. Imhoff, J. Kittle, A. S. Donigian, and R. C. Johansen. 1996. Hydrological Simulation Program—FORTRAN, user's manual for release 11. Athens, GA: U.S. Environmental Protection Agency, Environmental Research Laboratory.

Crawford, N. H., and R. K. Linsley. 1966. Digital simulation in hydrology: Stanford watershed model IV. Technical Report 39. Stanford, CA: Department of Civil Engineering, Stanford University.

Croley, T. E. II., 2002. Large basin runoff model. In *Mathematical models of large watershed hydrology*, ed. V. Singh, D. Frevert, and S. Meyer. Littleton, CO: Water Resources Publications, 717–770.

Croley, T. E., II, and C. He. 2005, Distributed-parameter large basin runoff model I: Model development. *Journal of Hydrologic Engineering* 10: 173–181.

Croley, T. E., II, and C. He. 2006. Watershed surface and subsurface spatial intraflows. *Journal of Hydrologic Engineering* 11(1):12–20.

Croley, T. E., II, C. He, and D. H. Lee. 2005. Distributed-parameter large basin runoff model II: Application. *Journal of Hydrologic Engineering* 10: 182–191.

Garen, D. C., and D. S. Moore. 2005. Curve number hydrology in water quality modeling: uses, abuses, and future directions. *Journal of the American Water Resources Association* 41: 377–388.

Ghadiri, H., and C. W. Rose. 1992, *Modeling chemical transport in soils: Natural and applied contaminants*. Ann Arbor, MI: Lewis Publishers.

Kawkins, R. H. 1978. Runoff curve number relationships with varying site moisture. *Journal of the Irrigation and Drainage Division* 104: 389–398.

Knisel, W. G. 1980. CREAMS: A fieldscale model for chemical, runoff, and erosion from agricultural management systems, Conservation Report No. 26. Washington, DC: U.S. Department of Agriculture, Science and Education Administration.

Lahlou, N., L. Shoemaker, S. Choudhury, R. Elmer, A. Hu, H. Manguerra, and A. Parker. 1998. *BASINS V.2.0 user's manual.* Washington, DC: U.S. Environmental Protection Agency Office of Water, EPA-823-B-98-006.

Leonard, R. A., W. G. Knisel, and D. A. Still. 1987. GLEAMS: Groundwater loading effects of agricultural management systems. *Transactions of ASAE* 30:1403–1418.

Sharpley, A. N., and J. R. Williams. 1990. *EPIC—erosion/productivity impact calculator,* Technical Bulletin No. 1768. Washington, DC: U.S. Department of Agriculture, Agricultural Research Service.

U.S. Environmental Protection Agency. 2002. *National water quality inventory 2000 report.* EPA-841-R-02-001. Washington, DC: U.S. Environmental Protection Agency.

Wischmeier, W. H., and D. D. Smith. 1978. *Predicting rainfall erosion losses.* Agricultural Handbook No. 537. Washington, DC: U.S. Department of Agriculture.

Young, R. A., C. A. Onstad, D. D. Bosch, and W. P. Anderson. 1989. AGNPS: A nonpoint-source pollution model for evaluating agricultural watersheds. *Journal of Soil and Water Conservation* 44:168–173.

Part III

Water Quality and
Biogeochemical Processes

10 Estimating Nonpoint Source Pollution Loadings in the Great Lakes Watersheds

Chansheng He and Thomas E. Croley II

10.1 INTRODUCTION

Nonpoint source pollution is the leading source of impairment of U.S. waters (U.S. Environmental Protection Agency [EPA] 2002). In the Great Lakes basin, contaminated sediments, urban runoff and combined sewer overflows (CSOs), and agriculture have been identified as the primary sources of impairments of the Great Lakes shoreline waters (U.S. EPA 2002). The problems caused by these pollutants include toxic and pathogen contamination of fisheries and wildlife, fish consumption advisories, drinking water closures, and recreational restrictions (U.S. EPA 2002). Management of these problems and rehabilitation of the impaired waters to fishable and swimmable states require identifying impaired waters that are unable to support fisheries and recreational activities and tracking sources of both point and nonpoint source material transport through a watershed by hydrological processes. Such sources include sediments, animal and human wastes, agricultural chemicals, nutrients, and industrial discharges, and so forth. While a number of simulation models have been developed to aid in the understanding and management of surface runoff, sediment, nutrient leaching, and pollutant transport processes such as ANSWERS (Areal Nonpoint Source Watershed Environment Simulation) (Beasley and Huggins 1980), CREAMS (Chemicals, Runoff and Erosion from Agricultural Management Systems) (Knisel 1980), GLEAMS (Groundwater Loading Effects of Agricultural Management Systems) (Leonard et al. 1987), AGNPS (Agricultural Nonpoint Source Pollution Model) (Young et al. 1989), EPIC (Erosion Productivity Impact Calculator) (Sharpley and Williams 1990), and SWAT (Soil and Water Assessment Tool) (Arnold et al. 1998), to name a few, these models are either empirically based, or spatially lumped, or do not consider nonpoint sources from animal manure and combined sewer overflows (CSOs) and infectious diseases. To meet this need, the National Oceanic and Atmospheric Administration (NOAA) Great Lakes Environmental Research Laboratory (GLERL) and Western Michigan University are jointly developing a spatially distributed, physically based watershed-scale water quality model to estimate movement of materials through both point and nonpoint sources in both surface and subsurface waters to the Great Lakes watersheds. The water quality model evolves from GLERL's distributed large basin runoff model (DLBRM) (Croley and He 2005; Croley et al. 2005). It consists of moisture storages of upper soil zone, lower soil zone, groundwater zone, and surface, which are arranged as a serial and parallel cascade of "tanks" to coincide with the perceived basin storage structure. Water enters the snowpack, which supplies the basin surface (degree-day snowmelt). Infiltration is proportional to this supply and to saturation of the upper soil zone (partial-area infiltration). Excess supply is surface runoff. Flows from all tanks are proportional to their amounts (linear-reservoir flows). Mass conservation applies for the snow pack and tanks; energy conservation applies to evapotranspiration. The model allows surface and subsurface flows to interact both with each other and with adjacent-cell surface and subsurface storages. Currently, it is being modified to add materials runoff through each of the storage tanks routing from upper stream downstream to the watershed outlet (for details of the model, see the companion paper by Croley and He 2006). This paper describes procedures for estimating potential loadings of sediments, animal manure, and agricultural chemicals into surface water from multiple databases. These estimates will be used as input to the water quality model to quantify the combined loadings of agricultural sediment, animal manure, and fertilizers and pesticides to Great Lakes waters for identifying the critical risk areas for implementation of water management programs.

10.2　STUDY AREA

The study area of this research is the Cass River watershed, a subwatershed of the Saginaw Bay watersheds. The Cass River watershed runs cross Huron, Sanilac, Tuscola, Lapeer, Genesee, and Saginaw counties of Michigan and joins the Saginaw River near Saginaw (Figure 10.1) and has a drainage area of 2,177 km^2. The Cass River is used for industrial water supply, agricultural production, warm-water fishing, and navigation. Agriculture and forests are the two major land uses/covers in the Cass River watershed, accounting for 60% and 21% of the total land area, respectively. Soils in the watershed consist mainly of loamy and silty clays and sands, and are poorly drained in much of the area. Major crops in the watershed include corn, soybeans, dry beans, and sugar beets. Over the years, the primary agricultural land use and associated runoff, improper manure management, poor municipal wastewater treatment, irrigation withdrawal, and channel dredging and straightening have led to high nutrient runoff, eutrophication, toxic contamination of fish, restrictions on fish consumption, loss of fish and wildlife habitat, and beach closures in the Cass River watershed (Michigan Department of Natural Resources 1988). Because of dominant agricultural land use and related high soil loss potential, the Cass River watershed was selected as the study area for estimating the loading potential of agricultural nonpoint sources to assist the management agencies in planning and managing NPS (nonpoint source) pollution control activities on a regional scale.

10.3　ESTIMATING SOIL EROSION POTENTIAL

Soil erosion is caused by raindrops, runoff, or wind detaching and carrying soil particles away. It is the most significant nonpoint source pollution factor affecting the

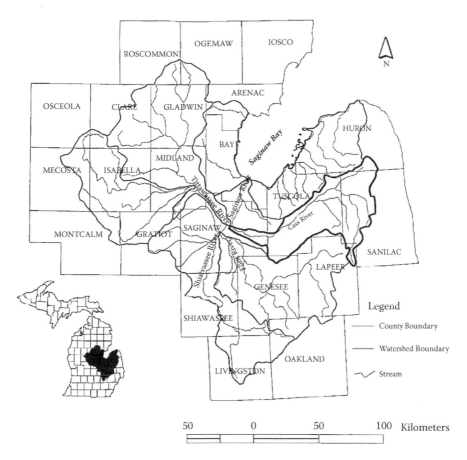

FIGURE 10.1 The Saginaw Bay watershed boundary.

quality of water resources in the United States. Soil erosion by water includes sheet and rill erosion. Sheet erosion is removal of a thin layer of soil from the surface of the land. Rill erosion is removal of soil from the sides and bottoms of small channels formed where surface runoff becomes concentrated and forms tiny streams. Sheet erosion and rill erosion usually occur together and are hence referred to as sheet-and-rill erosion (Beasley et al. 1984). Soil erosion by wind is the removal of soil by strong winds blowing across an unprotected soil surface. This study focuses on the potential of sheet-and-rill erosion by both water and wind at the watershed scale.

10.3.1 WATER EROSION POTENTIAL

The universal soil loss equation (USLE) (equation 10.1) is one of the most fundamental and widely used methods for estimating soil erosion and sediment loading on an annual basis (Wischmeier and Smith 1978). A number of simulation models, such as ANSWERS, EPIC, AGNPS, and SWAT, use the USLE for erosion and sediment simulation.

$$Y = R*K*LS*C*P*\text{Slope Shape} \qquad (10.1)$$

where Y is the computed average soil loss per unit area, expressed in tons/acre; R is the rainfall and runoff factor and is the rainfall erosion index (EI) plus a factor for runoff from snowmelt or applied water; K is the inherent erodibility of a particular soil; L is the slope length factor, S is the slope steepness factor; C is the cover and management factor; P is the support practice factor; and the slope shape factor represents the effect of slope shape on soil erosion (Wischmeier and Smith 1978, Young et al. 1989).

Realizing that the USLE is not intended for estimating erosion and sediment yield from a single storm event, we use AGNPS to estimate the soil erosion and sediment potential for illustration purposes since we have not incorporated the revised USLE (RUSLE) (Foster et al. 2000) into the distributed water quality model yet (He et al. 1993, 1994). AGNPS, based on the USLE, simulates runoff, erosion and sediment, and nutrient yields in surface runoff from a single storm event. Basic databases required for the AGNPS model include land use/land cover, topography, water features (lakes, rivers, and drains), soils, and watershed boundary (He et al. 1993; 1994; 2001; He 2003). The model output includes estimates of runoff volume (inches), sediment yield (tons), sediment generated within each cell (tons), mass of sediment attached and soluble nitrogen in runoff (lbs/acre), and mass of sediment attached and soluble phosphorus in runoff (lbs/acre).

The Digital Elevation Model (DEM) of 1:250,000 from the U.S. Geological Survey was used to derive slope and aspect. The STATSGO (State Soil Geographic Data Base) data from the U.S. Department of Agriculture Natural Resources Conservation Service were used to determine dominant texture, hydrologic group, and weighted soil erodibility. The 1979 land use/land cover data from the Michigan Resource Information System (MIRIS) and related hydrography databases were used to derive land use–related parameter values. The storm event chosen was a 24-hour precipitation of 3.7 inches with an average recurrence of 25 years. Fallow, straight-row crops, and moldboard plow tillage were assumed in the simulation.

The model was applied to the Cass River watershed with a spatial resolution of 125 ha (310 acres). (Note: the cell size was set at 310 acres to ensure the entire watershed was discretized to no more than 1,900 cells—the limit of AGNPS version 3.65.) The simulated results show that the runoff volume was higher in the agricultural land (Figure 10.2). The soil erosion rate simulated from the single storm event generally centered around 1 to 1.5 tons per acre, with no or little erosion in the forested areas and a greater rate (up to 5 tons per acre) in portions of the agricultural land. The sediment yield was highest (up to 45,000 tons in the 310-acre area) near the mouth of the watershed as the flatness of the area and lower peak runoff rate resulted in a higher rate of deposition. These results indicate that agricultural activity was a main nonpoint source pollution contributor under the worst management scenario (fallow, straight-row crops, and moldboard plow tillage) (He et al. 1993).

10.3.2 Wind Erosion Potential

Wind erosion results in more than five million metric tons of soil erosion, accounting for 63% of the total soil erosion in the Saginaw Bay watersheds (Michigan Department of Natural Resources 1988). The critical months for wind erosion are April and

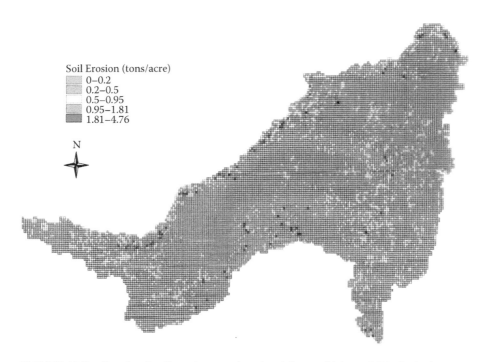

FIGURE 10.2   Simulated soil erosion rate (tons/acre) from a 24-hour, 3.7-inch single storm event in the Cass River watershed. (**See color insert after p. 162.**)

May in the Saginaw Bay basin. Few methods are available for estimating soil erosion by wind, such as the wind erosion equation developed by the U.S. Department of Agriculture (USDA), Agricultural Research Service Wind Erosion Laboratory (Woodruff and Siddoway 1965, Gregory 1984, Presson 1986). These methods are suitable for estimating wind erosion potential at the field level but difficult to use at the watershed level. As soil erodibility, wind, and quantity of vegetative cover are the main factors affecting wind erosion (Woodruff and Siddoway 1965), this study used soil association data and vegetation indices to estimate the wind erosion potential for the entire Cass River watershed.

STATSGO was used to extract six wind erodibility indices for all the soil associations in the Cass River watershed. These groups are (USDA Soil Conservation Service 1994):

Group 1: 310 ton/acre/year
Group 2: 134 ton/acre/year
Group 3: 86 ton/acre/year
Group 4: 56 ton/acre/year
Group 5: 48 ton/acre/year
Group 6: 38 ton/acre/year

These indices represent the wind erodibility based on the soil surface texture and percentage of aggregates.

TABLE 10.1

**Classification of wind erosion potential based
on the soil erodibility and NDVI values.**

	Classification criteria	
Wind erosion potential	NDVI	Wind erodibility group indices (tons/acre/year)
No wind erosion	>0.60	Any group indices (1–6)
Subtle wind erosion	0.40–0.60	134–300
Little wind erosion	0.40–0.60	>300
	or 0.20–0.39	or <140
Medium wind erosion	0.20–0.39	>140
	or 0.10–0.19	or <140
High wind erosion	0.10–0.19	>140
	or <0.10	or <100
Severe wind erosion	<0.10	>100

The LANDSAT 5 TM data of June 1, 1992 was used to derive the Normalized Differential Vegetation Indices (NDVI). These indices give a relative quantification of vegetation amount, with vegetated areas yielding high values, and nonvegetated areas yielding low or zero values. The formula for calculating the NDVI is:

$$NDVI = (TM\ Band\ 4 - TM\ Band\ 3) / (TM\ Band\ 4 + TM\ Band\ 3) \quad (10.2)$$

TM Bands 3 and 4 represent the red and near-infrared spectrum, respectively. The differential values between the two help us determine vegetation type, vigor, and biomass content (Lillesand and Kiefer 1987).

The wind erodibility group indices from STATSGO were combined with the NVDI to delineate the potential wind erosion areas. The criteria for classifying the wind erosion based on soil and vegetative factors are shown in Table 10.1.

Wind speed and direction were not considered in identifying the potential wind erosion areas because such variables were not available in the four second-order weather stations within or adjacent to the Cass River watershed. The closest first-order weather station that collects wind speed and direction data (Flint Weather Station) is about 50 miles south of the watershed. Soil moisture data was not considered in the delineation process because wind erosion occurs in the Saginaw Bay basin including the Cass River watershed in April and May when soil moisture is usually high in the region (Merva 1986).

The wind erodibility of the soil groups in the Cass River watershed ranged from 48 to 310 ton/acre/year based on the properties of soil associations from STATSGO. The NDVI derived from the LANDSAT TM data showed that about 33% of the Cass River watershed had NDVI value of between 0.01 and 0.20, 23% of the area with NDVI of 0.21 to 0.40, 39% of the land with NDVI value of 0.41 to 0.60, and about 6% of the land with dense vegetation cover (NDVI value of 0.61 to 1.00). As

TABLE 10.2

Distribution of wind erosion potential in the Cass River watershed.

Wind erosion potential	Acres	Percent
No wind erosion	257,756	44.3
Subtle wind erosion	57,069	9.8
Little wind erosion	120,131	20.7
Medium wind erosion	143,951	24.8
Severe wind erosion	2,155	0.4
Total	581,063	100.0

soil and vegetation are two of the most important factors affecting the wind erosion potential, the wind erodibility and NDVI were combined to produce a wind erosion map for the Cass River watershed. The results indicate that about 25% of the Cass River watershed had a medium wind erosion potential (Table 10.2) and most of the area was in the agricultural land.

10.4 ESTIMATING ANIMAL MANURE LOADING POTENTIAL

Improper management of animal manure can result in eutrophication of surface water and nitrate contamination of groundwater (He and Shi 1998). Differentiation of variations in soil and animal manure production within each county requires relevant data and information at a finer scale. The animal manure loading potential was estimated by using the 5-digit zip code from the 1987 Census of Agriculture (He and Shi 1998). Farm counts of animal units by 5-digit zip code were tabulated for cattle and hogs only in three classes: 0 to 49, 50 to 199, and 200 or more per zip code area (we used 49, 199, and 200 to represent the three classes of animals per zip code in our calculation). These data were matched with the 5-digit zip code boundary file and multiplied by animal manure production coefficients to estimate animal manure loading potential (tons/year) by zip code. The coefficients from the *Livestock Waste Facilities Handbook* MWPS-18 (Midwest Plan Service 1993) were used in this study: for a 1,000-lb dairy cow, annual manure (20%–25% solids content and 75%–80% percent moisture content) production: 13 metric tons, nitrogen 150 lbs, and phosphate 60 lbs; for a 150-lb pig, annual manure production: 1.6 metric tons, nitrogen 25 lbs, and phosphate 18 lbs. As the animal waste was likely applied to agricultural land, the loading potential was combined with agricultural land to derive the animal loading potential in tons per acre of agricultural land.

The results indicate that Huron and Sanilac counties produced the greatest animal waste loading potential per acre of land (over 30 tons per acre); Tuscola and Lapeer counties had the second highest loading potential (20–30 tons per acre) in the Cass River watershed (Figure 10.3). Portions of Sanilac and Tuscola counties had animal manure loading potential of over 40 tons per acre of land annually. Distribution of nitrogen and phosphate from animal manure by zip code shows a similar pattern. Huron, Sanilac, and Lapeer counties had the highest nitrogen and phosphate

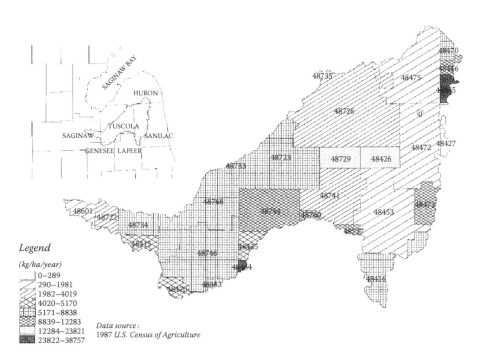

FIGURE 10.3   Distribution of animal manure (in kg/ha) by zip code in the Cass River water-shed. Data from U. S. Department of Agriculture, *1987 Census of Agriculture*, Washington, DC: U.S. Department of Agriculture, National Agricultural Statistics Service.

loading potential, Tuscola County had the second highest amount, and Saginaw and Genesee counties had the lowest loading potential in the Cass River watershed. At the zip code level, four zip code areas (48465, 48426, 48729, and 48464) had animal manure N production rates of greater than 150 lb/acre. Consequently, these locations can be targeted for implementation of manure management programs. This also indicates that agricultural statistics data at the finer scale (below county level) would reveal more useful information than would the county-level data in animal manure management. Large livestock operations difficult to identify at county level, could be easily identified using the 5-digit zip code level for manure management (He and Shi 1998).

The total loading potential for the animal manure, nitrogen, and phosphate was 10 million tons, 26 tons, and 21 tons, respectively, in the Cass River watershed, averaging 30 tons of animal waste, 160 lbs of nitrogen, and 130 lbs of phosphate per acre of agricultural land (Table 10.3). The high loading potential per acre of agricultural land makes optimal management of animal manure in the Cass River watershed necessary for minimizing the pollution potential to the surface and subsurface waters. These estimates, of course, do not include manure produced by other animals such as sheep and poultry. Thus, it is inevitable that discrepancies exist between the actual animal manure amount and these estimates. Users should realize the limitation of these estimates when using them for water resources planning.

TABLE 10.3

Estimated total amounts of animal waste, nitrogen, and phosphate from animal waste in the Cass River watershed based on the 1987 Census of Agriculture data.

Animal waste (tons)	Nitrogen (N) (tons)	Phosphate (P_2O_5) (tons)
9,632,000	25,700	21,180

10.5 AGRICULTURAL CHEMICAL LOADING POTENTIAL

Agricultural chemical data from the Michigan Department of Agriculture (MDA) were used to estimate the loading potential of agricultural chemicals (including both fertilizers and pesticides) in the Cass River watershed. The MDA Pesticide and Plant Pest Management Division (PPMD) maintains two databases for tracking pesticide use: (1) restricted-use pesticide (RUP) (pesticides that could cause environmental damage, even when used as directed), and sales-based estimates, which record all RUP sales in the state of Michigan; and (2) survey-based estimates, which provide estimates of pesticide use associated with each production type in a county by multiplying crop acreage by percentage of area treated and average application rates based on the 1990 and 1991 agricultural chemical usage survey data. Nitrogen fertilizer usage data were also estimated from the agricultural chemical usage survey data at the county level (USDA National Agricultural Statistics Services 1990, 1991). The uncertainty associated with the RUP sales-based estimates is that the locations of sales and applications of pesticides may not be the same. The problem with the survey-based estimates is that crop production estimates and pesticide application estimates are not available for all crops (Michigan Department of Agriculture, 1993). We used the survey-based estimates for pesticides and nitrogen fertilizer for estimating agricultural chemical loading potential in the Cass River watershed. The estimates were further adjusted by consulting the Michigan State University (MSU) Cooperative Extension Service pesticide expert (Renner, personal communication 1994). These estimates were lumped together to derive the average usage of pesticides per acre of cropland at the county level. They were not differentiated by their toxic level as this project focused on estimating the loading potential of total agricultural chemicals. Similarly, the usage of nitrogen fertilizers were divided by the total acreage of application cropland to derive the average usage of nitrogen fertilizer per acre of cropland. Average phosphate application data for all the cropland were based on the USDA National Agricultural Statistical Service's 1990 and 1991 field crops survey results at the state level (Table 10.4).

As shown in Table 10.4, approximately 15 million pounds of nitrogen and 13.5 million pounds of phosphate fertilizers, and 206,000 pounds of pesticides were applied to cropland in the Cass River watershed annually. Although these numbers represent the amounts applied to the crops and a major portion of these may be used by plants, some portions of these could be transported either through surface runoff

TABLE 10.4
Estimated agricultural chemical loading potential in the Cass River watershed.

County	Cropland (acres)	Nitrogen amount (lbs)	Phosphate amount (lbs)	Pesticides amount (lbs)
Huron	10,164	579,348	426,888	8,284
Genesee	4,932	247,093	207,144	3,290
Lapeer	11,713	585,650	491,946	7,567
Saginaw	21,729	956,076	912,618	15,428
Sanilac	136,114	6,669,586	5,716,788	83,030
Tuscola	137,389	6,182,505	5,770,338	88,341
Total	322,041	15,220,000	13,526,000	206,000

or drainage tiles to the surface waters or leached to groundwater in the watershed. Thus, implementing best management practices in applying agricultural chemicals is crucial for reducing the pollution potential in the Cass River watershed.

10.6 CRITICAL NONPOINT SOURCE POLLUTION AREAS

Taking into account the loading potential of soil erosion, animal manure, and agricultural chemicals, it seems that the overall nonpoint source pollution potential is highest in the Huron, Sanilac, and eastern Tuscola portions of the Cass River watershed. These areas, located in the upper stream of the Cass River watershed, are mainly cropland with relatively high slope and close proximity to drains and tributaries. The high fertilizer and pesticide application rate, and the large amount of animal manure from concentrated livestock industry, and intensive cropping activities make these areas a major source of potential contamination to the surface and subsurface waters in the Cass River watershed. In addition, as these areas are located in the upper stream, activities in these areas will have a greater impact on the water quality downstream.

The simulated sediment yield from AGNPS appears to be greatest in the mouth of the Cass River near Saginaw due to the flatness of the area. The larger amount of sedimentation in the area is likely to have a negative impact on aquatic habitat. It could also lead to elevated streambed and increased flooding frequency and damage in the surrounding areas. These areas could be targeted for future water quality management programs for minimizing nonpoint source contamination potential.

10.7 SUMMARY

The National Oceanic and Atmospheric Administration (NOAA) Great Lakes Environmental Research Laboratory (GLERL) and Western Michigan University are developing a spatially distributed, physically based watershed-scale water quality model to estimate movement of materials through point and nonpoint sources in both surface and subsurface waters to the Great Lakes watersheds. This paper, through a case study of the Cass River watershed, estimates loading potential of soil erosion and sediment by water and wind, animal manure and nutrients, and agricultural chemicals. The results suggest that the Cass River watershed introduces large amounts of nutrients and sediment into the Saginaw River and Bay. Soil erosion was up to 5 tons per acre in some agricultural land areas after a single 24-hour storm of 3.7 inches with frequency of one in 25 years. The sediment yield was up to 145 tons per acre at the outlet of the watershed. Total nitrogen and phosphorus runoff was higher in agricultural land. About 25% of the total land area in the Cass River watershed was subject to medium wind erosion. The concentrated animal industry produces approximately 10 million tons of manure, 26 tons of nitrogen, and 21 tons of phosphate in the Cass River watershed, averaging 30 tons of manure, 160 lbs of nitrogen, and 130 lbs of phosphate per acre of agricultural land annually. About 15 million lbs of nitrogen fertilizer, 13 million lbs of phosphate, and 206,000 lbs of pesticides were used annually in the agricultural land of the Cass River watershed. Portions of these fertilizers and pesticides could be transported, either through surface runoff or drainage tiles, to the streams and groundwater in the watershed.

Agricultural statistics data at the finer scale (below county level) would reveal more useful information than would the county-level data in estimating multiple sources of pollutant loading potential. Governmental agencies should consider collecting and tabulating relevant information at the township or zip code level to aid environmental planning and management.

This paper estimates the loading potential of multiple sources of pollutants but does not consider the pollutant transport through runoff processes in the entire watershed. Work is underway to provide these estimates to the distributed large basin runoff water quality model for simulating pollutant transport in both surface and subsurface water in the Saginaw Bay watersheds. Such information, once verified with the Saginaw Bay water quality data, will help management agencies and ecosystem researchers in prioritizing water quality control programs and protecting critical fisheries and wildlife habitat.

ACKNOWLEDGMENTS

This is GLERL Contribution No. 1376. Partial support from the Western Michigan University Department of Geography Lucia Harrison Endowment Fund and the Michigan State University Institute of Water Research is also appreciated.

REFERENCES

Arnold, G., R. Srinavasan, R. S. Muttiah, and J. R. Williams. 1998. Large area hydrologic modeling and assessment. Part I. Model development. *Journal of the American Water Resources Association* 34: 73–89.

Beasley, D. B., and L. F. Huggins. 1980. *ANSWERS (Areal nonpoint source watershed environment simulation)—user's manual*. West Lafayette, IN: Department of Agricultural Engineering, Purdue University,

Beasley, R. P., J. M. Gregory, and T. R. McCarty. 1984. *Erosion and sediment pollution control*. 2nd ed. Ames: Iowa State University Press.

Croley, T. E., II, and C. He. 2005. Distributed-parameter large basin runoff model I: Model development. *Journal of Hydrologic Engineering* (ASCE) 10:173–181.

Croley, T. E., II, C. He, and D. H. Lee. 2005. Distributed-parameter large basin runoff model II: Application. *Journal of Hydrologic Engineering* (ASCE) 10:182–191.

Croley, T. E., II, and C. He. 2006. Watershed surface and subsurface spatial intraflows. *Journal of Hydrologic Engineering* (ASCE) 11:12–20.

Foster, G. R., D. C. Yoder, D. K. McCool, G. A. Weesies, T. J. Toy, and L. E. Wagner. 2000. Improvements in science in RUSLE2. Paper No. 00-2147. ASAE, St. Joseph, MI.

Gregory, J. M.. 1984. Prediction of soil erosion by water and wind for various fractions of cover. *Transactions of the American Society of Agricultural Engineers* 27:1345–1350.

He, C. 2003. Integration of GIS and simulation model for watershed management. *Environmental Modeling and Software* 18:809–813.

He, C., and T. E. Croley II. 2005. Development of a distributed large basin operational hydrologic model. *Control Engineering Practice* (in review).

He, C., and T. E. Croley II. 2007 Application of a distributed large basin runoff model in the Great Lakes Basin. *Control Engineering Practice* 15:1001–1011.

He, C., J. F. Riggs, and Y. T. Kang. 1993. Integration of geographic information systems and a computer model to evaluate impacts of agricultural runoff on water quality. *Water Resources Bulletin* 29:891–900.

He, C., and C. Shi. 1998. A preliminary analysis of animal manure distribution in Michigan for nutrient utilization. *Journal of the American Water Resources Association* 34:1341–1354.

He, C., C. Shi, and Y. T. Kang. 1994. Modeling agricultural nonpoint source pollution potential in support of environmental decision making. East Lansing: Institute of Water Research, Michigan State University.

He, C., C. Shi, C. Yang, and B. P. Agosti. 2001. A Windows-based GIS-AGNPS interface. *Journal of the American Water Resources Association* 37:395–406.

Knisel, W. G., ed. 1980. CREAMS: A fieldscale model for chemical, runoff, and erosion from agricultural management systems, Conservation Report No. 26. Washington, DC: U.S. Department of Agriculture, Science and Education Administration.

Leonard, R. A., W. G. Knisel, and D. A. Still. 1987. GLEAMS: Groundwater loading effects of agricultural management systems. *Transactions of the American Society of Agricultural Engineers* 30:1403–1418.

Lillesand, T. M., and R. W. Kiefer. 1987. Remote sensing and image interpretation. New York: John Wiley & Sons.

Merva, G. E. 1986. *Bay County wind erosion research (final report).* East Lansing: Michigan State University, Department of Agricultural Engineering.

Michigan Department of Agriculture. 1991. *Michigan Agricultural Statistics 1990, 1991.* Lansing: Michigan Department of Agriculture.

Michigan Department of Agriculture. 1993. Differential mapping of pesticide use for Michigan's state management plan. Lansing: Michigan Department of Agriculture.

Michigan Department of Natural Resources. 1988. Remedial action plan for Saginaw River and Saginaw Bay. East Lansing: Michigan Department of Natural Resources, Surface Water Quality Division, Great Lakes and Environmental Assessment Section.

Midwest Plan Service. 1993. *Livestock Waste Facilities Handbook.* 3rd ed. MWPS-18, Midwest Plan Service, Ames: Iowa State University.

Presson, W. A. 1986. A wind erosion model to predict average soil loss for single events. MS thesis, Department of Agricultural Engineering, Michigan State University, East Lansing, Michigan.

Sharpley, A. N., and J. R. Williams, eds. 1990. *EPIC—Erosion/productivity impact calculator,* Technical Bulletin No. 1768. Washington, DC: U.S. Department of Agriculture, Agricultural Research Service.

U.S. Department of Agriculture, National Agricultural Statistical Service. 1990. *Agricultural chemical usage: 1990 field crops summary.* Washington, DC: U.S. Department of Agriculture.

U.S. Department of Agriculture, National Agricultural Statistical Service. 1991. *Agricultural chemical usage: 1991 field crops summary.* Washington, DC: U.S. Department of Agriculture.

U.S. Department of Agriculture, Soil Conservation Service. 1994. *State soil geographic data base (STATSGO) data users guide.* Miscellaneous Publication Number 1492, Washington, DC: U.S. Department of Agriculture.

U.S. Environmental Protection Agency. 2002. *National water quality inventory 2000 report.* EPA-841-R-02-001, Washington, DC: U.S. Environmental Protection Agency.

Wischmeier, W. H., and D. D. Smith. 1978. *Predicting rainfall erosion losses.* Agricultural Handbook No. 537, Washington, DC: U.S. Department of Agriculture.

Woodruff, N. P., and F. H. Siddoway. 1965. A wind erosion equation. *Soil Science Society of America Proc.* 29:602–608.

Young, R. A., C. A. Onstad, D. D. Bosch, and W. P. Anderson. 1989. AGNPS: A non-point-source pollution model for evaluating agricultural watersheds. *Journal of Soil and Water Conservation* 44:168–173.

11 Simulating Historical Variations of Nitrogenous and Phosphorous Nutrients in Honghu Lake Basin, China

Feng Gui, Ge Yu, and Geying Lai

11.1 INTRODUCTION

In recent years, eutrophication has become an increasingly serious environmental problem in lake systems. Excessive nutrient enrichment is the root cause of eutrophication. Although lakes naturally receive nutrient inputs from their catchments and the atmosphere, many human activities such as sewage inflows, runoff from agricultural fields, and industrial effluents have greatly accelerated the eutrophication process. To assess the relative roles of natural, climate-induced changes versus human-related activities, such as the removal of vegetation, it is important to evaluate the natural trajectory of nutrient transportation over the catchments and its contribution to a lake's eutrophication. The eutrophication of the lakes in the most developed region in China, the mid-lower reaches of the Yangtze River, has brought great attention from the public, scholars, and government. As there are many types of lakes in the Yangtze River basin, and the causes of eutrophication are different for each type, Honghu Lake basin, which is located at the middle reach of the Yangtze River, was chosen as the study area. Computational simulations with the SWAT (Soil and Water Assessment Tool) model were used to reflect the historical nutrient sedimentation and transportation processes. With the application of the SWAT model, the principle of nutrient transportation in the *natural agricultural environment* (the environment under which the nutrient sedimentation and transportation processes are only controlled by natural factors, such as topography, climate changes, and natural vegetation cover, etc.) has been discussed in order to provide scientific basis for the mechanism research of lake water eutrophication.

11.2 STUDY AREA

Honghu Lake, located at the middle reach of the Yangtze River (113°12'–113°26'E and 29°40'–29°58'N) (Figure 11.1), is typical of a shallow water lake. As the largest lake in Jianghan Plain, its geological setting is a faulted depression between the

129

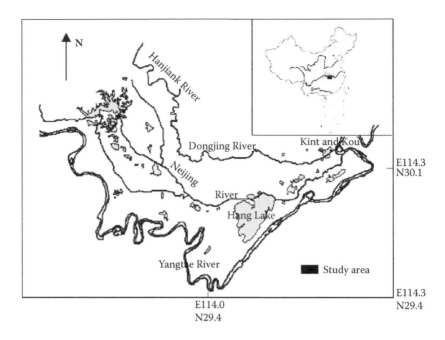

FIGURE 11.1   Location of study area.

Yangtze River basin and the Dongjing River basin. The Honghu Lake region covers both Honghu Lake County and Jianli County, an area of 344.4 km².[1,2] Due to the lake evolution and the impacts of human activities in recent centuries, Honghu Lake has shrunk significantly from its original coverage.[3,4]

Honghu Lake basin, located at the northernmost area of China's subtropical zone, features a typical northern subtropical humid monsoon climate, with abundant diurnal heating and precipitation in the same season. The average precipitation in the area is approximately 1100 to 1300 mm, 77% of which is in the summer season. The average annual runoff in this basin is approximately $37.35 \times 10^8 \, m^3$.[1,5] The area receives water from Chang Lake, San Lake, and White Dew Lake, with a watershed area of 8265 km². In the flood season, Honghu Lake receives water not only from the upstream area, but also from overflowing water from the downstream rivers. Prior to the 1960s, building of the floodgate of Xin Tankou, located downstream of Neijing River, the floodwater from the Yangtze River would overflow through the Neijing River to Honghu Lake (Figure 11.1). In the flood season, the basin area enlarges to 10,325 km².

11.3 INTRODUCTION OF THE SWAT MODEL

SWAT (Soil and Water Assessment Tool) is a river basin–scale hydrological model developed by the Agriculture Research Service of the U.S. Department of Agriculture.[6–8] Its GIS-based version (AVSWAT 2000) has strong functionality in spatial analysis and visualization. The SWAT model can be used to evaluate the impact of land management practices on water, sediment, and agricultural-chemical yields in large, complex basins with a variety of soils and land cover on a time scale of 10 to

100 years. There have been many successful applications of the SWAT model, with longtime series outputs to reconstruct the past environment or to predict future environmental changes.[9–11]

11.4 BOUNDARY CONDITIONS AND SIMULATION DESIGN

11.4.1 BOUNDARY CONDITIONS AND MODEL DATA PREPARATION

To simulate and evaluate the eutrophication of Honghu Lake basin under natural-agricultural environments, the boundary conditions were set as natural-agricultural time scenarios, including the following major factors: (1) natural topography, slope, and channel, (2) climatic and hydrological factors (temperature, solar radiation, precipitation, and runoff), and (3) biomass of natural vegetation.

Based on these boundary values, the following data sets were compiled or created for the model simulation: a database of land topography and hydrological units were generated based on the 1:250,000 digital elevation models (DEM) published by the National GIS Center of China.[12] A stream network data set was created by digitizing 1:100,000 topography maps and using a "burn-in" method[6] by which a stream network theme is superimposed onto the DEM to define the location of the stream network. This feature is useful in situations where the DEM does not provide enough detail to allow the interface to accurately predict the location of the stream network.[6] A 1:1,000,000 digital soil map with spatial and nonspatial information was obtained from the Institute of Soil Science, Chinese Academy of Science (ISSCAS),[13] including information on sand, silt, organic material, soil pH value, total phosphorus, available phosphorus, and bulk density. Due to the difference in the soil texture system between the SWAT standard (U.S. standard) and the standard of the second national soil survey in China, the dataset used in this study was transformed into the SWAT standard. Soil bulk density, available water capacity of the soil layer (SOL_AWC), and saturated hydraulic conductivity (SOL_K) were calculated using SPAW 6.1.[14] The soil types were further classified into four different hydrological groups (A, B, C, and D) according to the guidelines documented in the SWAT user manual.[8]

Land use/vegetation data, obtained from a 1:1,000,000 China vegetation map,[15] were transformed to a grid file with a resolution of 10'x10' to correspond to the plant nutrient data. In this study, the vegetation data were used to provide land use information. Within the study area, the major vegetation types are rice field and wetland. Thus, the corresponding land use types were classified into rice land, water, and forested wetland, forested mixed.

Meteorological data were collected from weather stations located within or near the basin, including daily precipitation, maximum and minimum temperature data from 1951 to 2000, as well as daily radiation, average wind speed, and humidity from 1980 to 2000. To generate meteorological data over the last 200 years, statistical values (such as standard deviation, skew coefficient of the daily precipitation and temperature in a month, and the probability of a wet day after a dry day, etc.) reflecting the characteristics of the local climate were calculated by using the weather generator tool in the SWAT model.

All data mentioned above were transformed into the Arc/INFO grid format with an Albers equal-area projection.

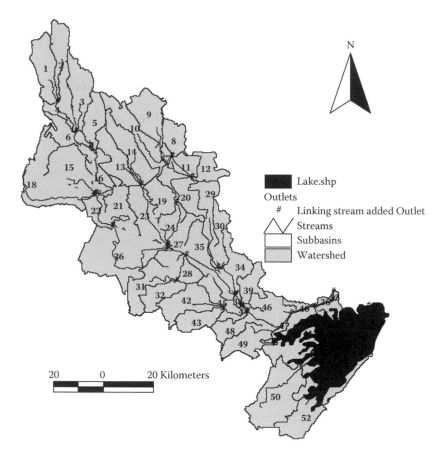

FIGURE 11.2   Watershed delineation and subbasin division for non-flood period.

11.4.2 SIMULATION PROCEDURE AND DESIGN

For the simulation experiment, we designed two scenarios according to the difference in watershed areas between the flood season (from June through August) and the non-flood season (from March through May as well as from September through February) (Figures 11.2 and 11.3).

We handle this source area change by adjusting the watershed outlet location as well as the number. We selected one outlet location at the entrance of Honghu Lake for the winter, while for summer we chose an outlet downstream of Honghu Lake.

Our procedure for simulation began with the construction of a background and dynamic database of each subbasin, including geological sediment, topography, climate, hydrology, soil, and vegetation coverage. We then subdivided the watershed into several subbasins. Subbasins possess a geographic position in the watershed and are spatially related to each other. We likewise subdivided the subbasins into HRU (hydrologic response units). HRUs are portions of a subbasin that possess unique land use/management/soil attributes. To acquire the function of the

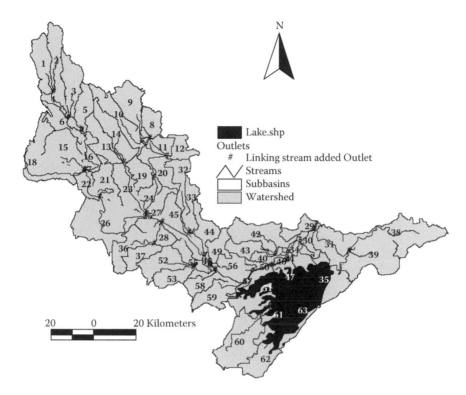

FIGURE 11.3   Watershed delineation and subbasin division for flood period.

nutrient source and transport dynamic of the land surface process, we evaluated all of the databases to the subbasin and HRU levels. The next step involved running the model with a time step of 24 hours and a continuous simulation of 200 years. We then used sedimentary core records to compare and validate the simulation output.

11.5 RESULTS AND DISCUSSION

11.5.1 SIMULATION OUTPUT ANALYSIS

The source area change has little effect on the nutrient concentration, but has a larger effect on the nutrient production and flow flux. There is an increase of approximately 25% in the flow flux and nutrient production in the summer season experiment.

11.5.1.1 Variability and Characteristics of Input Flow Flux

The results showed that the average yearly runoff flux was $46.1 \times 108 \text{ m}^3$. The runoff distribution within a year has its peak value during the period from April through September, in which the summer season (from June to August) contributes approximately one-half of the entire year's runoff (Figure 11.4a).

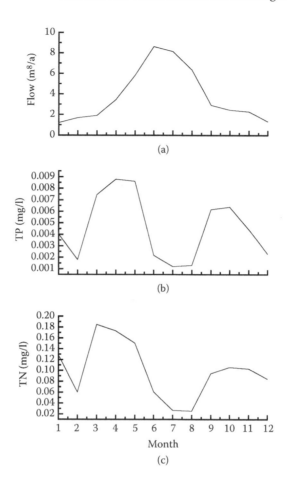

FIGURE 11.4   Simulation of annual mean of input flow flux, TN, TP.

11.5.1.2 Nutrient Changes in a Year

The annual means of TN (total nitrogen) and TP (total phosphorus) were plotted in Figure 11.4. A negative correlation was observed between the flow flux and the concentration of TN and TP, a lag between the peak value of flow and nutrient concentration. Due to the confluence of a channel in a large subbasin, only a portion of the surface runoff will reach the main channel on the day it is generated, creating a lag between the time the surface runoff was generated and the time it reaches the main channel.

In the summer season when the rains are heaviest, the nutrient concentration reached its lowest value because the nutrients were diluted by an abundant flow flux. The highest annual nutrient concentration occurred in spring, perhaps due to cultivation activities during that period of time (Figures 11.4b, c).

11.5.1.3 Variations of Nutrient Concentration over Time

The nutrient concentration changes through time were analyzed seasonally. Both TN and TP concentrations exhibited high variability. The maximum TP concentration appeared in the spring season and the minimum appeared in the summer, with a relatively stable concentration in the time series. The maximum TN concentration appeared in the spring and the minimum appeared in summer, with a slow increasing trend in the time series (Figure 11.5).

11.5.1.4 Annual TP and TN Production

The annual average TN and TP production of the Honghu Lake basin was 420.25 tons per year and 19.613 tons per year, respectively. Analysis of the nutrient production showed that the production of TN had a slow increasing trend with time, while TP had no obvious trend. The nutrients in the Honghu Lake basin are characteristic of an accumulated natural trend (Figure 11.6).

11.5.2 VALIDATIONS OF SIMULATION OUTPUTS

It is difficult to directly validate the 100-year simulations of nutrient production and concentration due to the lack of long-term observation data. However, the simulations can be indirectly validated by comparing the data with long-term sedimentary nutrient records and establishing the statistical relations between them. In this study, we focused on the study results from the 84-cm-long sedimentary cores HN (for Honghu Lake North; collected at the northern part of Honghu Lake in November 2002 with a water depth of 3.2 m), and the relative ^{137}Cs-dating data, with the 150-cm-long core H2-2002 (collected at the central part of Honghu Lake in 2002) as the reference.[16,17] The analyses indicated that the sedimentation rate of Honghu Lake in the last 540 years was about 0.155 cm/a. At 25 ~ 8 cm, the age approximately corresponds to the years 1840 to 1950.[16,17] Above 8 cm, the age corresponds to the year 1950. Chen Ping et al. (2004) determined the sedimentation rate of the core H2-2002 to be approximately 0.092 ~ 0.129 cm/a at layer D (above 22 cm in the core), suggesting an age of approximately 150 years, from 1845 to 1992.[16] The dates of the two cores are almost the same in each layer.

Because the simulation covers the time period between 1840 and 1950, a natural-agricultural time, the simulated outputs could be validated by comparing the TN and TP concentrations of relative layers with these age data of the cores as follows:

1. The nutrient concentrations in the core HN varied (Figure 11.7). Between 1840 and 1850, the TN concentrations increased while core depth decreased and the concentration variability was 1.20 ~ 1.77 g/kg. Between 1959 and 2002, the nutrient concentration increased rapidly with the decrease of the depth, and the variability was 1.77 ~ 8.78 g/kg. The TP concentration varied through the core profile, with a peak value of 65 ~ 70 cm at the topmost 4 cm. The TP concentration also showed an increasing trend with the decrease of depth, and reached 0.946 g/kg at the depth of 0.25 cm (Table 11.1).

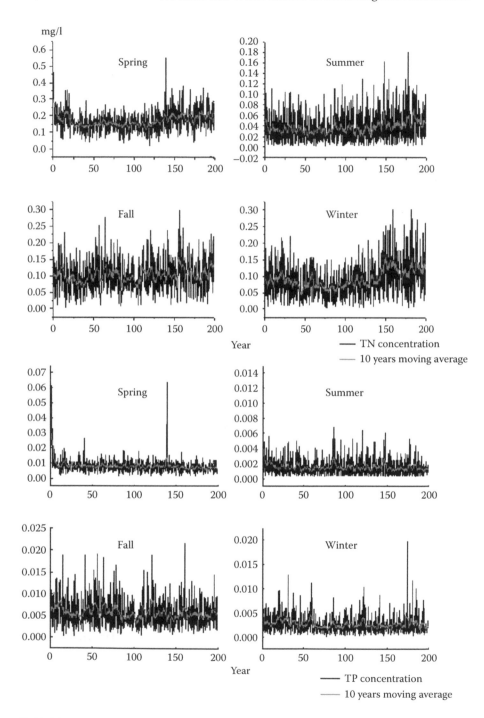

FIGURE 11.5   Change of nutrient concentration with time.

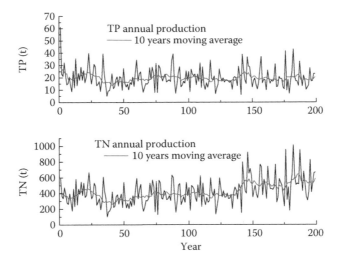

FIGURE 11.6   Simulations of annual total phosphorus and total nitrogen production.

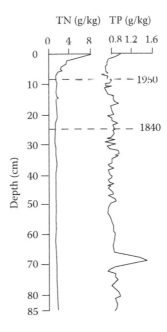

FIGURE 11.7   Total nitogen and total phosphorus values in core HN (From Yao Shuchun, Bin Xue, and Weilan Xia. Human impact recorded on the sediment of Honghu Lake. *Journal of Hohai University* (Natural Sciences), 2004, 32 (Supplement): 154–159.

TABLE 11.1

The water nitrogen concentrations statistically calculated based on the sedimentary nitrogen values.

	Core TN (g/kg)	Calculated nitrogen in water (mg/l)
1960s	2	0.28
1970s	3	0.32
1980s	3.8	0.41
1990s	5.8	0.558
2002	8.3	0.597

2. The recent observations of water nutrients were obtained from the literature[3,4] and the measured data (the measured TN and TP concentrations of the Honghu Lake water). Table 11.1 shows the water nutrient information. We compared it with sedimentary records since the 1950s in order to examine the relations between the water nutrients and the sedimentary nutrients. A regression was obtained as follows:

$$Y = 0.2475 \text{Ln}(X) + 0.0864 \tag{11.1}$$

where Y is the total nitrogen of the three types in water (mg/l) and X is the TN concentration in the sedimentary core (g/kg).

The correlation coefficient $R^2 = 0.9557$ was significant for the sample size. With this statistical relation, the sedimentary nutrient was used to calculate the water nutrient in a time series, which was then applied to validate the results of simulated nutrients in the lake water.

As shown in Table 11.2, during the period of 1840 to 1950, the variability of nitrogen in water was 0.13 ~ 0.23 mg/l, with an average value of 0.19 mg/l. The simulated nitrogen concentration for the same period was 0.071 ~ 0.11 mg/l, with an average value of 0.09 mg/l.

According to Table 11.2, the difference between the simulations and referenced calculations is approximately 50%. This difference may have occurred because there are two sources of lake nutrients, the internal (lake) source and the external source, while the simulation only took into account the nitrogen from the external sources. As such, the simulation output represents the level of the nutrients transported from the entire basin into the lake water and does not account for the nutrients produced from the lake water itself. As many studies [18,19] have suggested that the internally sourced nutrients play an important role in lake eutrophication, these nutrients should be considered in order to fully understand nutrient level changes in lake water. While the contribution of internally sourced nutrient release in Honghu Lake cannot presently

TABLE 11.2
Comparisons of simulated nitrogen concentrations with calculated values.

	Simulated TN (mg/l)	Calculated nitrogen values (mg/l)	Accuracy
Maximum	0.114	0.228	0.502
Average	0.090	0.195	0.463
Minimum	0.076	0.132	0.581

be estimated, it should not be neglected, and considering this effect, our simulation results could be better compared.

11.6 CONCLUSIONS

The simulations using the SWAT model helped explain the nutrient changes in the Honghu Lake basin. Both the nutrient concentrations in sediment records and the water body showed an increasing trend since the 1950s. The nutrients have slowly accumulated in the basin; the increase in the past 50 years appeared more significant than the 50 years prior to that.

The study indicated that the SWAT model could be an effective tool for evaluating the nutrient production and changes in a lake watershed on a time scale of several hundreds of years. Thus, the model simulation may help understand some effects of long-term climate changes under human impacts.

ACKNOWLEDGMENTS

This study was funded by the National Key Fundamental Research and Development Planning Program (Grant No. 2002CB412300-1) and by the Innovation Program of Nanjing Institute of Geography and Limnology, Chinese Academy of Sciences (Grant No. KZCX1-SW-12) and all measured data were provided by Yang Xiangdong.

REFERENCES

1. Wang, Suming, and Hongshen Dou, eds. 1998. *China Lake*. Beijing: Science Press, 580.
2. Compile committee of water conservancy. 2000. *Hubei Water Conservancy Records* Beijing: China Water Conservancy and Water Electricity Press.
3. Chen, Shijian. 2001. Environmental problems and ecological countermeasures for Honghu Lake in Hubei province. *Journal of Central China Normal University* (Natural Sciences) 35(1):107–110.
4. Wang, Fei, and Qiming Xie. 1990. The succession trend and management countermeasure for Hong wetland ecosystem. *Rural Eco-environment* 2:21–25.
5. Xiang Guorong, ed. 1997. *Research on sustainable development of wetland agriculture in Sihu region*. Beijing: Science Press, 271.

6. Di Luzio, M., R. Srinivasan, J. G. Arnold, et al. 2002. *Arcview interface for SWAT2000 user's guide.* College Station: Texas Water Resources Institute TR-193, 345.
7. Neitsch, S. L., J. G. Arnold, J. R. Kiniry, et al. 2002. *Soil and Water Assessment Tool theoretical documentation, version 2000.* College Station: Texas Water Resources Institute TR-193, 498.
8. Neitsch, S. L., J. G. Arnold, J. R. Kiniry, et al. 2002. *Soil and Water Assessment Tool user's manual, version 2000.* College Station: Texas Water Resources Institute TR-193, 438.
9. Hu, Yuanan, Shengtong Cheng, and Haifeng Jia. 2003. Hydrologic simulation in NPS models: Case of application of SWAT in Luxi watershed. *Research of Environment Science* 16(5):29–32.
10. Jayakrishnan, J., et al. 2005. Advances in the application of the SWAT model for water resources management. *Hydrological Processes* 19:749–762.
11. Borah, D. K., and M. Bera. Watershed-scale hydrologic and nonpoint-source pollution models: Review of applications. *American Society of Agricultural Engineers* 47(3):789–803.
12. National GIS Center of China. 1999. Database of 1:2500,000 topography of China. Beijing: National GIS Center of China.
13. Shi, X. Z., D. S. Yu, E. D. Warner, et al. 2004. Soil database of 1:1,000,000 digital soil survey and reference system of the Chinese Genetic Soil Classification System. *Soil Survey Horizon* 45(4):129–136.
14. Saxton, K. *Soil—Plant—Atmosphere—Water (SPAW) field and pond hydrology operational manual (version 6.02).* Washington, DC: U.S. Department of Agriculture, Agricultural Research Service, 23.
15. Hou, Xueyu. 1988. *China physical geography (botanical geography).* Beijing: Science Press, 318.
16. Chen, Ping, Baoyan He, Eudo Kunihiko, et al. 2004. Records of human activities in sediments from Honghu Lake. *Journal of Lake Sciences* 16(3):233–237.
17. Yao, Shuchun, Bin Xue, and Weilan Xia. 2004. Human impact recorded on the sediment of Honghu Lake. *Journal of Hohai University* (Natural Sciences) 32 (Supplement):154–159.
18. Wetzel, R. G. 2001. *Limnology—lake and river ecosystem.* London: Academic Press.
19. Qin, Boqiang, and Chengxin Fan. 2002. Exploration of conceptual model of nutrient release from inner source in large shallow lakes. *China Environment Science* 22(2):150–153.

12 Predictive Modeling of Lake Nitrogen, Phosphorus, and Sediment Concentrations Based on Land Use/Land Cover Type and Pattern

Pariwate Varnakovida, Narumon Wiangwang,
Joseph P. Messina, and Jiaguo Qi

12.1 INTRODUCTION

In watershed management and planning, one of the major problems in lakes is the need to reduce nonpoint source pollution.[1] Specific land use and land cover (LULC) types, such as "cropland" and "urban", are associated with human activities and their physical characteristics often affect water quality.[2] Land use practices and water resources are unequivocally linked. The type and the intensity of land use have a strong influence on the receiving water resource, especially in lakes.[3]

Since agricultural land is dominant in many watersheds of the upper Midwest of the United States, it is considered a leading source for nonpoint source pollutants, primarily sediments and nutrients. Agricultural erosion occurs when fields are cleared of vegetation to prepare for crop planting. The physical erosion potential of some soil types, such as fine sandy loam, may be exacerbated by nonconservation agricultural practices, which may reduce the soil's chemical fertility. The angle and length of slopes on the land also influence the rate and amount of runoff, and in turn influence erosion. As soil fertility declines, farmers tend to increase their application of fertilizers. This intensive application of fertilizer then becomes a major source of culturally driven eutrophication in lakes and streams.[4–6]

Urbanization is another factor that impacts water quality. Catchment or subwatershed increases in impervious areas cause a direct impact on stream quality.[7] An impervious area is any area that no longer allows rainfall to soak into the ground, such as roads, sidewalks, rooftops, and driveways. When a site is developed, it loses its natural storage potential for rainfall. Consequently, rain that previously infiltrated into the ground evaporates or transpires, and rain that was temporarily stored in depressions and tree canopies now rapidly runs off the site.

Previous studies on lake eutrophication have focused either on external loadings or internal processes. Studies of external sources are generally limited to a single source and cannot be easily generalized.[8] Hence, to be able to provide insight into the effects of all the potential pollutant sources it is necessary to model the entire ecosystem leading to the in-lake effects. System modeling and simulation is one of the best alternatives to urban lake monitoring and water quality studies.

Although several simulation models are currently being used among researchers to predict water quality (e.g., BASINS [Better Assessment Science Integrating Point and Non-Point Sources], AGNPS [Agricultural Non-Point Source], and WASP [Water Quality Analysis Simulation Program]), most are either scale dependent or have extensive input data requirements. Therefore, for this study, we constructed the NPSSIM (nitrogen, phosphorus, and sediment simulation) model to predict TN (total nitrogen), TP (total phosporus), and TSS (total sediment) concentration in lakes based on surrounding LULC types and patterns with the express desire to manage scale and data problems, such as the interaction among local and regional processes and the lack of adequate data for model calibration.[9]

Most other models simulate output based on only one continuous landscape (e.g., a watershed or a region). NPSSIM differs from other models in three main ways. First, the simulation is based on a lakeshed scale. Only the lake and the surrounding land that drain into it are used in the model. The model treats lakes as a single entity; therefore, lakes from different locations, such as lakes in Michigan and lakes in Wisconsin, are no different. Nutrients and sediment from different lakes across landscapes can be simulated and compared in one step. Second, the model uses landscape metrics as the main approach. Landscape metrics are known to be able to describe characteristics of land use/land cover. It is also capable of comparing LULC from different locations through the landscape neutral model. NPSSIM is therefore truly spatial scale independent because as long as DEM (digital elevation model), LULC, soil type, and rainfall data are available on the same spatial resolution (e.g., 30 m for LANDSAT TM+ and 30 m NEDDEM), the model can simulate nutrient concentration regardless of the lake location. Even if spatial resolutions of the data are inconsistent, grain size adjustment is uncomplicated. Third, NPSSIM requires fewer input data than most other distributed watershed models. Although the model requires fewer inputs, it certainly incorporates hydrologic processes by using rainfall, evaporation, and ground permeability (e.g., soil type) parameters.

This model advances the state of knowledge of landscape pattern metrics as factors for the prediction of water quality parameters. Landscape indices are generally widely used tools for spatial landscape analyses and serve as standards for comparison between landscapes in different parts of the world. Linking water quality to landscape indices allows us to compare land and waterscapes across scales. The NPSSIM model serves to enhance our understanding of the relationships between lake water quality parameters and the wider landscape components, while providing water quality managers with a new cost-effective tool to manage water resources.

12.2 METHODS

12.2.1 LAND USE/LAND COVER (LULC) DATA

LULC data obtained from the Michigan Geographic Data Library were derived from a classification of 2001 Landsat Thematic Mapper (TM) imagery from three seasons: spring (leaf-off), summer, and fall (senescence).

12.2.2 DIGITAL ELEVATION MODEL (DEM)

A seamless 30-meter resolution DEM was produced by the United States Geological Survey (USGS). The data were checked for artificial sinks that may have been caused by the production process. All sinks smaller than 8 pixels were filled. A raster grid of geographically corrected water bodies, including both streams and lakes, were overlaid on the DEM. A focal filter was used to smooth out the elevation and eliminate sinks (Figure 12.1).

12.2.3 LAKE SAMPLING METHOD

The lake water quality parameter data used in this study were collected by the Michigan State University research group of Dr. Robert Jan Stevenson in the Department of Zoology. The data was comprised of data from 158 lakes within the Muskegon River watershed. The water quality parameter data were collected in spring and

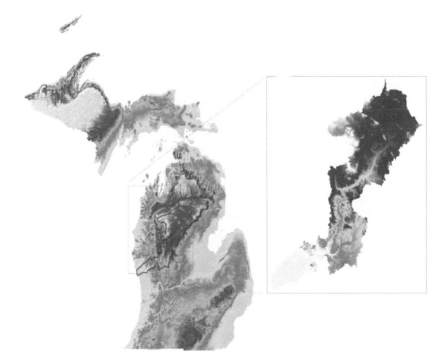

FIGURE 12.1   Digital elevation model (DEM): Michigan and the Muskegon River watershed. **(See color insert after p. 162.)**

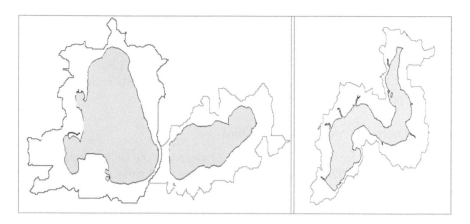

FIGURE 12.2 Example of lakeshed boundary: Lake Mitchell (left), Lake Cadillac (middle), and Hardy Dam pond (right).

summer of 2002. The samples were ideally collected from the deepest basin of the lake. One sample was collected from each lake to represent the lake. Lake data were classified by trophic status to ensure variation in water quality from oligotrophic to hypereutrophic. Due to the lack of a completed water quality parameter dataset, 52 lakes were selected from the database to be used in model simulation. Thirty percent of the simulated dataset (17 lakes) were used for the validation.

12.2.4 LAKESHED GENERATION

The lakeshed, commonly known as catchments of selected lakes, were generated using ArcINFO GRID software. To derive lakesheds, direction grids and source grids were needed. The direction grid was prepared by running the *flowdirection command* on the DEM. The source grid is a grid that represents cells above which the contributing area was prepared using a *streamlink function*. Streamlink assigns unique values to sections of a raster linear network between intersections. Once the direction grid and the source grid were completed, the lakeshed was determined by using a *watershed function*. The lakesheds were converted to the vector data model (Figure 12.2), and boundaries were edited when necessary. Next, lakeshed boundaries were used to clip out the surrounding LULC and were classed to seven major categories: urban, agriculture, open land, water, golf course, forest, and wetland.

12.2.5 LANDSCAPE PATTERN METRICS

In many studies, landscape pattern metrics have been used to describe changes in a landscape through time or to compare landscapes.[10] In this research, metrics were calculated on the LULC data including (1) number of patches per class per year, (2) total area per class per year, (3) mean patch size per class per year, (4) patch size standard deviation per class per year, and (5) dominance. All were calculated using the interactive data language (IDL) (TM).

12.2.6 Relationships among Landscape Metrics and Water Quality Parameters

The landscape pattern metrics and slope were regressed against TP, TN, and TSS in SYSTAT 9.0 with water quality parameters (TP, TN, and TSS) being the dependent variables and the landscape parameters (landscape pattern metrics and slope) the independent variables. Metrics that produced insignificant results ($P > 0.05$; confidence level 0.95) were eliminated. Finally, combinations of metrics that suggested the strongest relationships with nutrients and turbidity were used in the model. Mean patch size per year for classes including urban, agriculture, and open land were used.

12.2.7 Simulation Model

The NPSSIM predictive model was developed in the STELLA 8.0 software. TN, TP, and TSS were predicted in runoff using the regression relationship between landscape and the water quality parameters (as described in section 12.2.6). Other considerable factors, such as soil permeability, rainfall, lake volume, fertilizer application rate, nutrients concentration in rainfall, nutrients outflow rates, and TSS precipitate rate, were incorporated into the STELLA conceptual framework (as shown in Figure 12.3) to account for the hydrological process. Model constant and equations were based on regression results and Michigan Agricultural Statistics.

12.2.8 Model Verification and Validation

After model equations were revised, and adjusting coefficients were applied where needed, it was verified by comparing simulation results with field observations. Due to data limitations, the model was validated with TN, TP, and TSS data from 17 lakes (equivalent to 30 percent of all obtainable lake data) from the same year. These lakes were not used to create the model. Landscape metrics of the lake catchments were calculated (as described in sections 12.2.4 and 12.2.5) and entered into the model. TN, TP, and TSS were simulated in NPSSIM. The results were correlated with the in situ concentrations of the same lakes. Correlation coefficients (r^2) were used to report accuracy of the model.

12.2.9 Sensitivity Analysis

Sensitivity analysis was calculated on TN and TP in response to changes in mean patch size (MPS) of agriculture and urban changes. TN was more sensitive to agriculture when the patch size increased. Phosphorus was less sensitive to urban changes when the patch size increased. The analysis of sensitivity on fertilizer application and rainfall was performed; however, the results were not significant. Changes in fertilizer and rainfall up to 50% of the original rate did not have a significant impact on the water quality according to the model.

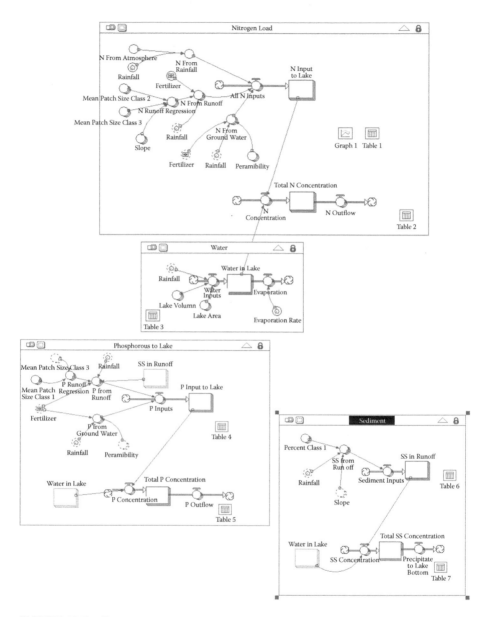

FIGURE 12.3 Conceptual nitrogen, phosphorus, and sediment simulation (NPSSIM) model.

12.3 RESULTS

A. Multivariate regression analysis suggested that the mean patch size of agriculture and open land in the catchments had the strongest relationship with TN, implying that increasing sizes of agriculture patches release more N into lakes. The result supported the hypothesis that TN increased as agriculture increased.

B. The regression result suggested that the mean patch size of urban and open land had the strongest relationship with TP. The result did not support the hypothesis that TP would increase as agriculture increased. Urban land use seemed to have more impact on TP than agriculture. This may be because a major portion of phosphorus transport in water is by binding with other substrates. Sediment loads increase with urban land use. That, combined with the open land that creates sediment loads in runoff, could enhance phosphorus transportation into the lakes.

C. TSS was most affected by the percentage of the area of urban land use and the landscape slope. The regression result supported the hypothesis that TSS increased as urban area increased. Urban built-up material created impervious surfaces, which increased velocities in surface runoff. Higher slopes also enhanced the process. Clearly increasing speeds in surface runoff increases water carrying capacity for particles and sediment.

D. Validation of the model between observed and predicted TP, TN, and TSS resulted in r^2 of 0.4535, 0.2471, and 0.4593, respectively (Figure 12.4).

12.4 CONCLUSION AND DISCUSSION

Lake ecosystems are very complicated and direct relationships among TN, TP, TSS, and water quality parameters are extremely difficult to determine. Correlation coefficients from our multivariable regression models were low ($r^2 < 0.25$); however, with some other variables, such as slope, soil permeability, rainfall, nutrients concentration in rainfall, and fertilizer included into the model, the prediction accuracy improved (r^2 0.25–0.46). Nonlinear or stepwise regression may capture the changes and improve the predictability. A stochastic model, as opposed to our deterministic NPSSIM model, may be an alternative approach because many variables (e.g., rainfall, fertilizer application) are not constant in nature.

Even though the validation correlation coefficient did not show impressive accuracy, the nutrient models did capture the trend of how water quality parameters respond to changes in LULC (bar graphs in Figure 12.4). The TN increased as agricultural areas increased as a result of the intensive application of fertilizer use. The TP increases as a catchment or subwatershed increases in impervious area with a direct impact on stream quality increases. The TSS model seemed to predict the average in all lakes. However, the percentage of urban area was selected as the significant variable in the regression process, which shows that the urban area was more correlated to TSS than other LULC types. Velocities in surface runoff were increased by impervious surfaces in urban land. On the other hand, area that had been classified as agricultural land was mostly covered with vegetation; therefore, a smaller amount of sediment was washed out of the surface. Additional parameters may need to be included in the model to enhance the response of TSS to differences in LULC. Sensitivity analysis shows an insignificant relationship between water quality parameters and fertilizer application. This was unexpected. The model may lack variables that better link fertilizer to the entire system. Further revision is needed.

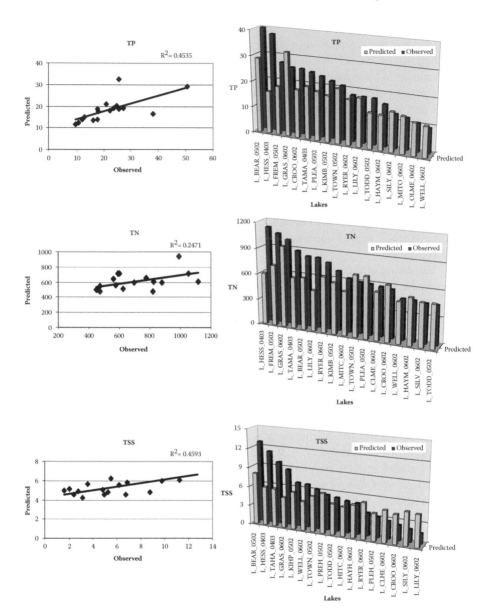

FIGURE 12.4   Validation results.

Without a time constraint, the model could potentially be improved. Refined equations and constants for the model could be developed. The input parameters were based on available data and may have reduced model predictability. This project and model serve as preliminary work for future research.

REFERENCES

1. Copeland, C. 1999. *Clean water action plan: Background and early implementation.* Washington, DC: CRS Report, 98–150.
2. Bianchi, M., and T. Harter. 2002. *Nonpoint sources of pollution in irrigated agriculture.* Farm water quality planning reference sheet 9.1. Oakland: University of California, 1–8.
3. Institute of Water Research. Michigan State University. http://www.iwr.msu.edu/edmodule/water/luwhome.htm.
4. Carpenter, S. R., N. F. Caraco, D. L. Correll, R. W. Howarth, A. N. Sharpley, and V. H. Smith. 1998. Nonpoint source pollution of surface waters with phosphorus and nitrogen. *Ecological Applications* 8(3):559–568.
5. Tinker, P. B. 1997. The environmental implications of intensified land use in developing countries. *Philosophical Transactions: Biological Sciences* 352(1356):1023–1032.
6. Mueller, D. K., and D. R. Helsel. 1996. *Nutrients in the nation's waters—too much of a good thing?* U.S. Geological Survey Circular 1136. Washington, DC: U.S. Geological Survey.
7. Roa-Espinosa, A., J. M. Norman, T. B. Wilson, and K. Johnson. 2003. Predicting the impact of urban development on stream temperature using a Thermal Urban Runoff Model (TURM). In *Proceedings of the National Conference on Urban Stormwater: Enhancing Programs at the Local Level, Chicago, Illinois, February 17–20, 2003.* Washington, DC: U.S. Environmental Protection Agency, 369–389.
8. Puijenbroek, P. J. T. M., J. H. Janse, and J. M. Knoop. 2004. Integrated modeling for nutrient loading and ecology of lakes in the Netherlands. *Ecological Modelling* 174:127 141.
9. Shanahan, P., M. Henze, L. Koncsos, W. Rauch, P. Reichert, L. Somlyody, and P. Vanrolleghem. 1998. River water quality modelling: II. Problems of the Art. *Water Science and Technology* 38(11):245–-252.
10. Iverson, L. R. 1988. Land-use changes in Illinois, USA: The influence of landscape attributes on current and historic land use. *Landscape Ecology* 2:45–62.

Part IV

Wetland Biology and Ecology

13 Soil Erosion Assessment Using Universal Soil Loss Equation (USLE) and Spatial Technologies— A Case Study at Xiushui Watershed, China

Hui Li, Xiaoling Chen, Liqiao Tian, and Zhongyi Wu

13.1 INTRODUCTION

Accelerated soil erosion is one of the most serious environmental problems in the world. In China, millions of tons of topsoil are eroded and transported every year, which not only degrades soil resources but also causes detrimental environmental consequences. Soil erosion affects productivity by changing soil properties, and particularly by destroying topsoil structure, reducing soil volume and water holding capacity, reducing infiltration, increasing runoff and washing away nutrients such as nitrogen, phosphorus, and organic matter (Meyer et al. 1985; Oyedele 1996). The resulting sediments act as carriers of pollutants including heavy metals, pesticides, and others.

Jiangxi is a province that suffers severely from soil erosion. The total affected area is 336.12×10^4 ha, which accounts for 95.5% of the total provincial area, and is mainly distributed in the upper and middle valley of the Xiu River, Ganjiang River, Xin River, Fu River, and around Poyang Lake.

The Xiushui watershed discharges water and sediments into Poyang Lake, which is the largest freshwater lake in China and an important international wetland with considerable ecosystem functions. Regional economic development, deforestation, and soil erosion in the Xiushui watershed have degraded the wetland ecological environment of Poyang Lake. Before effective management measures can be taken, the amount and location of soil that has been eroded must be quantified.

There are many models available for erosion estimation. Some of these models are based on physical parameters such as the WEPP (Water Erosion Prediction Project), and some are empirically orientated, such as the universal soil loss equation

153

(USLE). However, modeling soil erosion is difficult because of the complexity of the interactions of factors that influence the erosion (Wischmeier and Smith 1978). The objective of this paper is to estimate soil erosion and prioritize watersheds with respect to the intensity of soil erosion using the USLE.

13.2 STUDY AREA

Xiushui watershed is a subset of the Poyang Lake watershed in Jiangxi Province (Figure 13.1). It covers 14,606 km^2 and is located between 28° 22′29″ to 29° 32′18″ north latitude and 114° 3′15″ to 115° 55′32″ east longitude. Most of the watershed is mountainous area ranging from about 1 m to 1772 m above sea level with an average elevation of 341 m above sea level. The Xiu River runs from the southwest to the east and then discharges into Poyang Lake. The watershed is characterized by a fragile

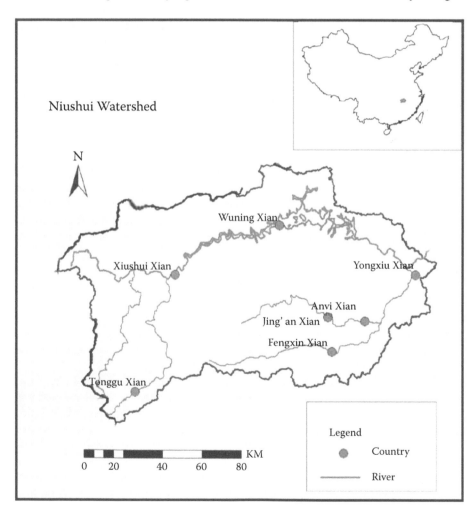

FIGURE 13.1 Location of Xiushui watershed.

ecosystem with frequent floods and relatively lagged development compared with its neighborhood, due to its unique geographic characteristics.

The watershed is situated in a subtropical zone with a monsoonal climate. The annual average temperature is 17°C. Annual precipitation averages 1613.7 mm, of which 73.1% occurs between March and August. The dominant agricultural crops are rice, cotton, and tea. The major soil types consist of red soil, brown soil, yellow-brown soil, weakly developed red soil, yellow soil, and paddy soil. The land is partially cultivated while the rest is covered with vegetation.

13.3 METHODS

The overall methodology involves using a soil erosion model, USLE, in a GIS (geographic information system) that incorporates data derived from remote sensing imagery, statistical data obtained from weather stations, and information from soil surveys. Individual raster data layers were built for each factor in USLE and processed by cell-grid modeling procedures in GIS to account for the spatial variability across the domain. With a consideration of the resolutions of all source data and the study site, the grid cells were set to 100 × 100 square meters.

13.3.1 Governing Equation

The USLE was hailed as one of the most significant developments in soil and water conservation in the twentieth century. It is an empirical technology that has been applied around the world to estimate soil erosion by raindrop impact and surface runoff. The USLE provides a quick approach to estimating long-term average annual soil loss. The model was originally developed and widely applied for a plane area. However, studies in mountainous areas have been conducted as well, and the results verified its ability to model complex landscapes (Bancy et al. 2000, Lufafa et al. 2003). It is expressed as follows:

$$A = R \cdot K \cdot L \cdot S \cdot C \cdot P \qquad (13.1)$$

where A is annual soil loss (t ha^{-1} yr^{-1}); R is the rainfall erosivity factor; K is the soil erodibility factor; L is the slope length factor; S is the slope steepness factor; C is the crop and management factor; and P is the conservation supporting practices factor. L, S, C, and P are dimensionless.

13.3.2 Determining the USLE Factor Values

13.3.2.1 Rainfall Erosivity (R) Factor

The R factor represents the rainfall and runoff's impact on soil. Originally, it was calculated as the total kinetic energy of the storm and its maximum 30-minute intensity (I30). Frequently, however, there are not enough data available to compute the R value using this method, especially for a large area. Different replacement methods have been developed over time for the computation of R. An erosivity index for river basins, developed by Fournier (1960), was subsequently modified by the FAO (Food and Agriculture Organization of the United Nations) as follows:

$$F = \sum_{i=1}^{12} r_i^2 / P \tag{13.2}$$

where r_i is the rainfall per month and P is the annual rainfall. This index is summed for the whole year and found to be linearly correlated with the EI30 index (R) of the USLE as follows:

$$R = a \cdot F + b \tag{13.3}$$

where a and b are the constants that need to be determined and vary widely among different climatic zones. You and Li (1999) presented the values of a and b for Taihe County, Jiangxi province, which is only one hundred kilometers away from the study area.

According to his study, a and b are 4.17 and −152, respectively. The unit of R was then converted into MJ mm ha^{-2} h^{-1}. Due to the large area of the watershed, data from seven meteorological stations were chosen to calculate the precipitation of the entire watershed. Among the seven stations, one is situated within the watershed, and the other six are in the neighborhood of the study area. Monthly rainfall data of seven stations over a time span from 1971 to 2000 were collected from the national meteorological bureau. The R value was calculated based on each of the seven stations by using the aforementioned method, and then interpolated into a continuous surface in GIS.

13.3.2.2 Soil Erodibility (K) Factor

The K factor measures soil susceptibility to rill and inter-rill erosion. Various methods for computing the K value were developed by researchers. As for this study, the detailed soil properties such as silt, sand, clay, and organic matter content could be acquired from the results of China's second soil survey. Liang et al. (1999) studied the area's soil erodibility and presented the K factor values corresponding to different soil types. In this study, we adopted their results for the estimation.

13.3.2.3 Topographic Factor (LS)

Slope length and slope gradient have substantial effects on soil erosion by water. The two effects are represented in the USLE by the slope length factor (L) and the slope steepness factor (S). L and S are best determined by pacing or measuring in the field, but extensive fieldwork is both time consuming and labor extensive. A digital elevation model (DEM) is a useful source for describing the topography of the land surface and is employed in LS calculation. There are some problems found in LS estimation by traditional methods, which assume that the length factor is defined as the distance to the divide or upslope border of the field. However, two-dimensional overland flow and the resulting soil loss actually depend on the area per unit of contour length contributing runoff to that point. The latter may differ considerably from the manually measured slope length, as it is strongly affected by flow convergence and/or divergence (Desmet and Govers 1996). The new concept was forwarded and some software such as Usle2D (Desmet and Govers 2000) was designed to overcome this problem by replacing the slope length by the unit contributing area.

13.3.2.4 Crop and Management Factor (C)

The C factor in the USLE measures the combined effect of the interrelated cover and crop management variables (Folly et al. 1996). The C factor could be evaluated from long-term experiments where soil loss is measured from land under various crops and crop management practices. However, such experimental installations are rarely available for a wide range of areas. Remote sensing provides a powerful tool for the observation and study of landscapes. Vegetation indices (VI) are robust spectral measures of the amount of vegetation present on the ground. They typically involve transformations of spectral information to enhance the vegetation signal and allow for precise intercomparisons of spatiotemporal variations in terrestrial photosynthetic activity (United States Geological Survey [USGS] 2004). Vegetation indices (VI) are widely used to measure the amount, structure, and condition of vegetation. Evidence indicates that there is a relationship between the VI and C factor (Tweddale et al. 2000). With this in mind, we could develop a more efficient method for C factor estimation. Ma (2003) and Cai et al. (2000) presented the relationship between vegetation cover and NDVI (Normalized Distance Vegetation Index), vegetation cover and C factor, respectively. They are expressed as follows:

$$V_c = 108.49 I_c + 0.717 \qquad R^2 = 0.77 \qquad (13.4)$$

where V_c is vegetation cover (%) and I_c is the NDVI.

The following is the relationship between C factor and vegetation cover:

$$\begin{cases} C = 1 & V_c \leq 0 \\ C = 0.658 - 0.3436 \lg V_c & 0 < V_c \leq 78.3\% \\ C = 0 & V_c \geq 78.3\% \end{cases} \qquad (13.5)$$

where C is the C factor in the USLE. MODIS Level 3 series products cover NDVI, and the USGS NDVI data used in this study was compiled based on the images obtained from June 1 to 15, 2004.

13.3.2.5 Erosion Control Practice Factor (P)

The erosion control practice factor (P factor) is defined as the ratio of soil loss with a given surface condition to soil loss with up-and-downhill plowing. The P factor accounts for the erosion control effectiveness of such land treatments as contouring, compacting, establishing sediment basins, and other control structures (Angimaa et al. 2003). However, most of the study areas are mountains covered with forest, and there is no significant conservation practice installed. In this study, P was assumed to be 1.

13.4 RESULTS AND DISCUSSION

13.4.1 FACTORS IN USLE

The monthly average rainfall and the calculated rainfall erosivity are listed in Table 13.1, which shows that most of the precipitation was concentrated in May, June, and July. This result suggests that most of the erosion might occur within the rainfall season and can be largely ascribed to major storms.

The rainfall erosivity ranges from 5,733.4 to 12,628 and the highest erosivity was observed in Lushan, which is situated just northeast of the watershed. The nearby Jiujiang station has an erosivity of only 5,733.4 for the lower elevation with less rainfall compared to Lushan. Nanchang, the northernmost station with the most adequate rainfall, has an erosivity of 9,284.1. Jian, whose station is latitudinally located between Xiushui and Nanchang, has less rainfall erosivity compared to Nanchang. The general rainfall erosivity is shown in Figure 13.2.

TABLE 13.1
Monthly average of rainfall and rainfall runoff erosivity for each meteorological station.[a]

Station	Jan	Feb	Mar	Apr	May	Jun	Jul	Aug	Sep	Oct	Nov	Dec	Erosivity
Xiushui	70.2	93.6	147.9	222.9	215.4	299.4	177.9	116.7	84.6	78.9	63.6	42.6	8520.8
Lushan	75.9	99.6	157.5	224.1	258.0	315.9	249.9	289.2	149.1	115.5	85.5	48.0	12628
Nangchang	74.1	100.8	175.5	223.8	243.9	306.6	144.0	129.0	68.7	59.7	56.7	41.4	9284.1
Pingjiang	72.9	89.4	146.1	198.0	214.2	251.7	174.3	134.7	73.2	76.8	60.9	39.9	7162
Jian	73.4	103.2	169.0	224.4	214.6	234.0	116.3	134.5	79.6	74.2	55.0	40.7	7041.8
Jiujiang	51.8	95.0	137.0	183.6	193.1	213.7	141.0	131.8	95.5	96.5	64.8	40.3	5733.4
Jiayu	58.5	73.2	124.5	166.3	188.3	244.8	163.0	123.6	75.1	95.7	64.4	36.8	6017.3

[a] Units for rainfall and erosivity are mm and MJ mm hm^{-2} h^{-1}, respectively.

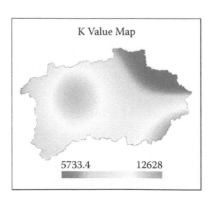

FIGURE 13.2 Map of rainfall erosivity. (**See color insert after p. 162.**)

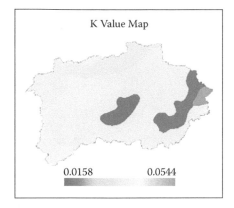

FIGURE 13.3 Map of soil erodiblility. **(See color insert after p. 162.)**

TABLE 13.2
K values for major soils.[a]

Soil	Red earth	Brown earths	Yellow-brown earths	Weakly developed red earths	Yellow earths	Moisture paddy
K value	0.0304	0.0158	0.0288	0.0299	0.0252	0.0544

[a] Units for soil erodibility is MghMJ^{-1}mm^{-1}.

The K factor value for each soil type was obtained from previous studies done in the area. The K factor map was thus prepared by assigning the K value to each soil type in a soil map. The values are given in Table 13.2 and the map shown in Figure 13.3. The erodibility of soils in this area varied from 0.12 for brown soil to 0.413 for moisture paddy soil. As shown in the K value map (Figure 13.3), the most easily erodible soil is only distributed in the easternmost portion of the watershed and covers a very small area. The soil with the biggest erosion is in the middle and eastern part of the study area and did not account for the larger area as well. The rest of the watershed is occupied by soils with relatively moderate erodibility.

The LS factor was calculated from the DEM for the entire watershed (Figure 13.4). The statistics demonstrate the variation of LS values (Table 13.3). We can determine from the LS map that the low LS value (flat area) is distributed along the valleys of the Xiushui River and its tributaries. The high LS value is in the mountainous area with steep slopes, which may result in higher amounts of erosion. The LS value ranges from 0 in very flat valleys to more than 300 in steep mountains. As to the distribution of LS values, 37.31% of the area is under 10, which indicates that the region is not topographically prone to erosion. LS values between 10 and 50 account for 37.51% of the watershed. The rest exhibit high LS values of more than 50 and extremely high values of more than 300, which cover 24.86% and 0.33%, respectively, and will surely result in severe erosion if no conservation practices are installed. Such large

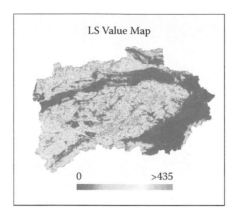

FIGURE 13.4   Map of topography. **(See color insert after p. 162.)**

TABLE 13.3
LS **distribution for the watershed.**

LS	Cell counts	Percent (%)	LS	Cell counts	Percent (%)
0–10	546781	37.31%	50–100	252131	17.20%
10–20	181687	12.40%	100–200	98130	6.70%
20–30	146752	10.01%	200–300	14025	0.96%
30–50	221265	15.10%	>300	4849	0.33%

variation of LS values can be ascribed to the complex mountainous landforms of the area, which is very typical in the erosion-stricken areas of southern China.

A map of cover and management factors is shown in Figure 13.5. It could be generally concluded that most of the watershed area is well covered with dense vegetation except certain sites in the northern and southern mountains whose severe deforestation would result in a very high C value and thus might lead to serious erosion.

In this study, a grid cell size (of all raster layers) was set to 100×100 m. However, the original resolution of the DEM is 93×93 m, and MODIS NDVI's is 250×250 m. The nearest neighborhood resample method was used to transform the raster layers into the desired resolution with an accuracy of less than one pixel. Given the same resolutions, the raster layers could be conducted using GIS overlay procedures.

The resolution will affect the accuracy of the result. The finer the resolution, the better the accuracy yields and vice versa. However, the fine resolution increases the amount of data, which results in longer processing time and the need for greater storage capacity. It is usually suitable for detailed analysis in small geographic areas. The coarse resolution has no such problems but it leads to larger errors. Taking both study area and input efforts into consideration, we identified the resolution to be 100×100 m, which was found to be appropriate and effective.

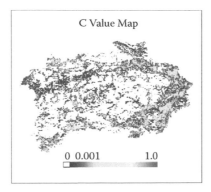

FIGURE 13.5   Map of cover and management. **(See color insert after p. 162.)**

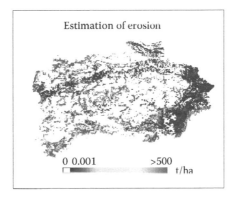

FIGURE 13.6   Map of erosion intensity. **(See color insert after p. 162.)**

13.4.2 EROSION INTENSITY

After the factor values were assigned or calculated for each of the grid cells, the factor maps were overlaid to produce a visualization of soil erosion estimation (Figure 13.6). The map indicates that the whole area is generally at very low risk for erosion.

Some statistical results showed that annual average soil losses for the watershed were 14.36 tons/ha and the standard deviation was 27.28 tons/ha, which suggests that the variation among estimations for the entire watershed was rather small. However, some extremely high estimations of more than 500 tons/ha occur in certain places, which is in accord with the current situation as mentioned in the introduction section of this paper. Measures, such as constructing terraces, strip cropping and returning field to forest should be taken to prevent further soil erosion.

The estimation was further prioritized into six classes: very slight, slight, moderate, severe, very severe, and extremely severe, according to the soil erosion classification criterion of China (Figure 13.7). From Figure 13.7, we can conclude that

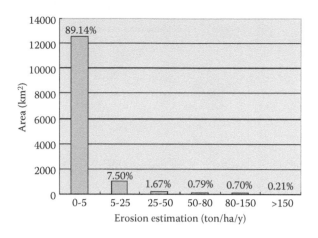

FIGURE 13.7   Histogram of erosion estimation.

89.14% of the watershed is under the tolerable erosion amount (5 tons/ha); 10.86% of the study area undergoes erosion, among which only 0.7% and 0.21% suffer from very and extremely severe erosion, respectively. Some very high estimates were observed in mountainous places with bad deforestation and could be distributed into the high *LS* and *C* values for these places. The rest of the watershed is relatively less affected by erosion.

As seen in the maps, the estimated erosion is very sensitive to the *LS* and *C* factors. The patterns in *LS* and *C* value maps are very similar to those of the erosion map, which may illustrate again that the soil conservation measures should be aimed at decreasing slope with less length and providing better cover to protect soil from rainfall and runoff detachment.

This method is not verified by real data for there is no measured data available. However, a four-day intensive field measurement effort was made in early July 2005 in order to collect ground truth information for erosion intensity. Thirty-three sites were checked and the vegetation cover and slope were investigated to estimate the erosion level. According to the field analysis, the estimations of this method generally reflected the erosion conditions of this watershed. Further investigations were made to explain the most likely reasons for the erosion, which could be summarized as follows: the construction new roads, the construction of quarries, the chopping of the forest for fuel or wood, and forest fires. All of the activities result in poor vegetation cover, thus exposing the soil directly to raindrop splash and runoff detachment.

13.5 CONCLUSIONS

In general, it is clear from the results of this study that USLE is an effective model for the qualitative as well as quantitative assessments of soil erosion intensity for the purposes of conservation management. Remote sensing imaging has provided valuable data sources, and the MODIS Level 3 VI products provide robust vegetation measurements for derivation of the *C* factor in this study. It is difficult to estimate the

FIGURE 2.3

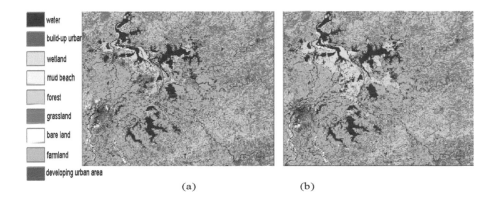

water	
build-up urban	
wetland	
mud beach	
forest	
grassland	
bare land	
farmland	
developing urban area	

(a)   (b)

FIGURE 2.4

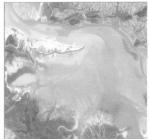

FIGURE 2.5

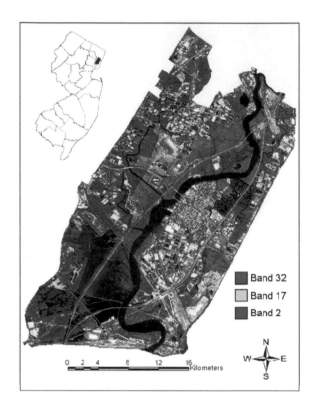

FIGURE 3.1

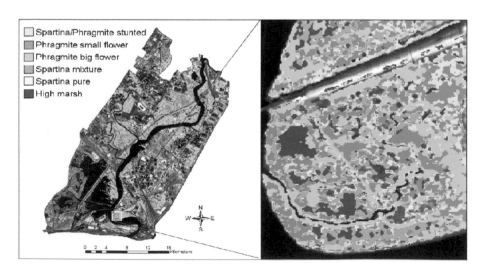

FIGURE 3.3

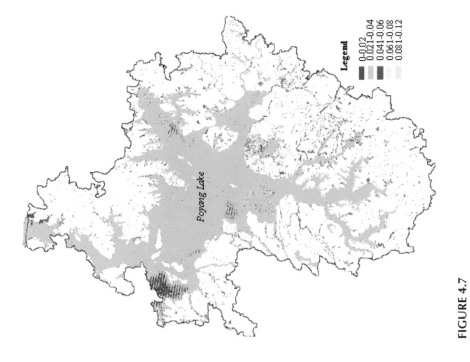

Poyang Lake

FIGURE 4.7

Legend
- 0-0.02
- 0.021-0.04
- 0.041-0.06
- 0.061-0.08
- 0.081-0.12

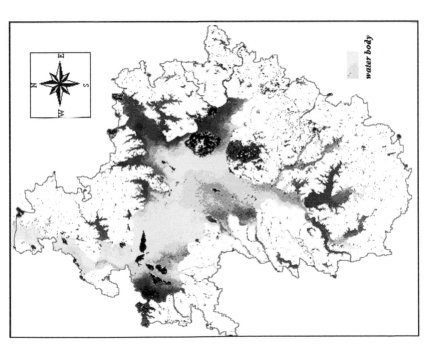

water body

FIGURE 4.6

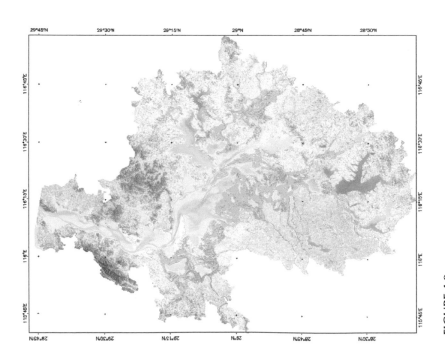

Legend
■ Field sites
▨ C. chinerascens

FIGURE 4.11

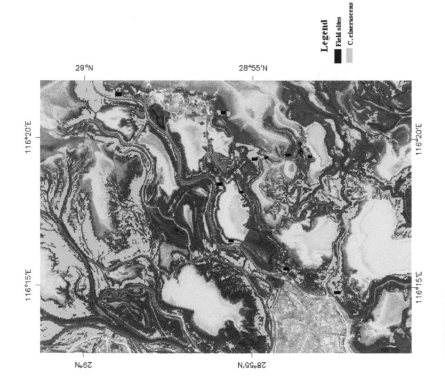

FIGURE 4.9

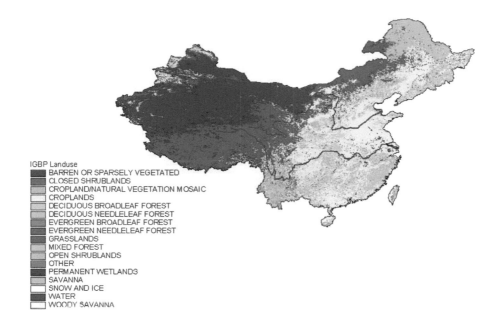

IGBP Landuse
BARREN OR SPARSELY VEGETATED
CLOSED SHRUBLANDS
CROPLAND/NATURAL VEGETATION MOSAIC
CROPLANDS
DECIDUOUS BROADLEAF FOREST
DECIDUOUS NEEDLELEAF FOREST
EVERGREEN BROADLEAF FOREST
EVERGREEN NEEDLELEAF FOREST
GRASSLANDS
MIXED FOREST
OPEN SHRUBLANDS
OTHER
PERMANENT WETLANDS
SAVANNA
SNOW AND ICE
WATER
WOODY SAVANNA

FIGURE 7.1

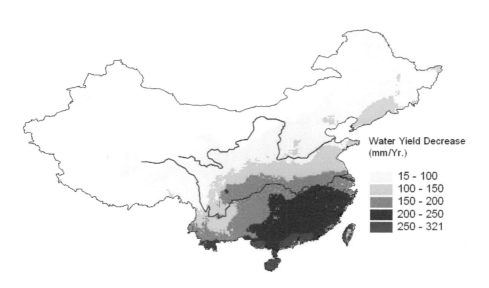

Water Yield Decrease
(mm/Yr.)

15 - 100
100 - 150
150 - 200
200 - 250
250 - 321

FIGURE 7.5

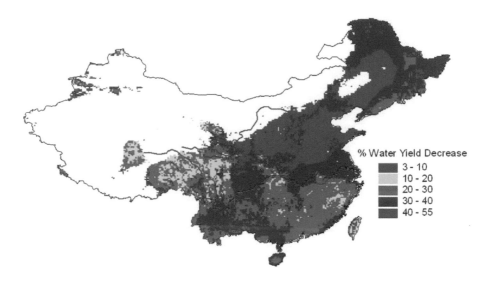

% Water Yield Decrease
- 3 - 10
- 10 - 20
- 20 - 30
- 30 - 40
- 40 - 55

FIGURE 7.6

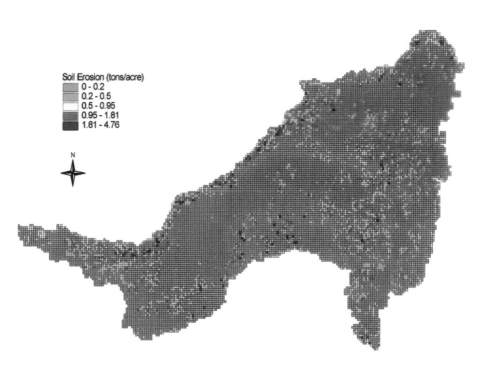

Soil Erosion (tons/acre)
- 0 - 0.2
- 0.2 - 0.5
- 0.5 - 0.95
- 0.95 - 1.81
- 1.81 - 4.76

N

FIGURE 10.2

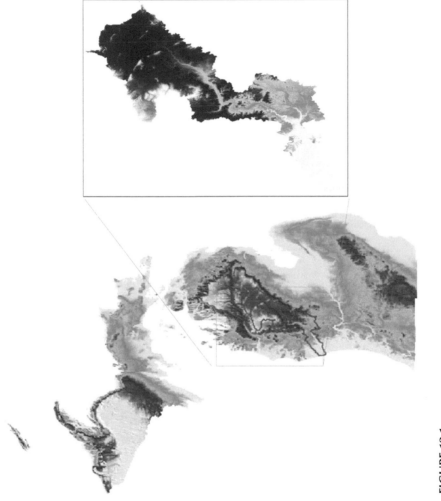

FIGURE 12.1

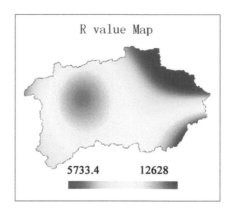

FIGURE 13.2

FIGURE 13.3

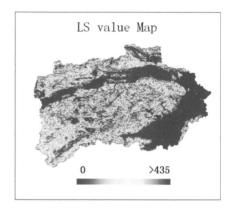

FIGURE 13.4

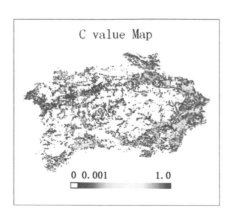

FIGURE 13.5

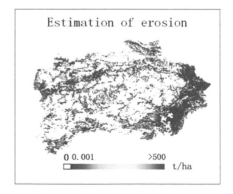

FIGURE 13.6

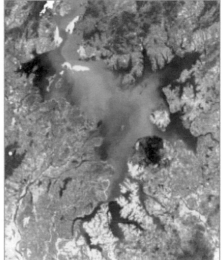

FIGURE 16.1

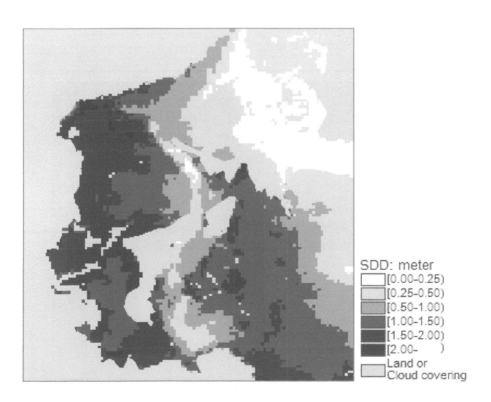

SDD: meter
☐ [0.00–0.25)
[0.25–0.50)
[0.50–1.00)
[1.00–1.50)
[1.50–2.00)
[2.00–   )
☐ Land or
  Cloud covering

FIGURE 16.6

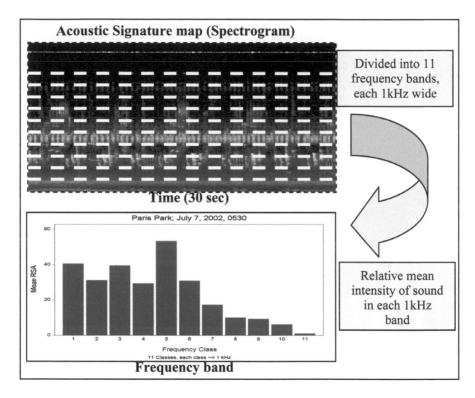

FIGURE 17.2

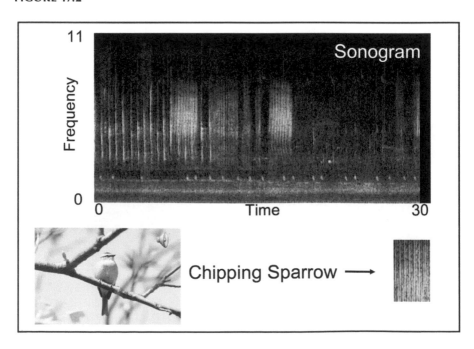

FIGURE 17.7

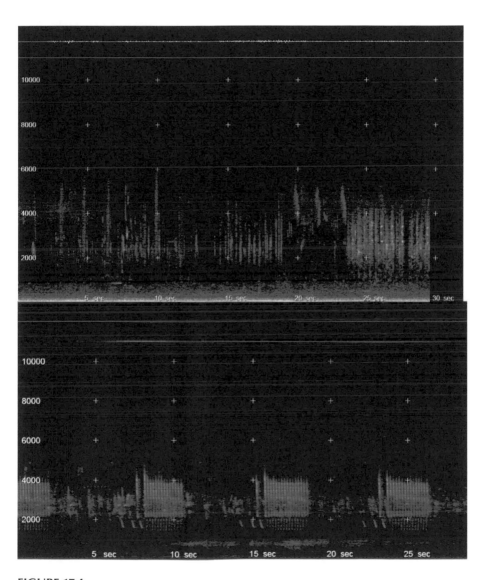

FIGURE 17.4

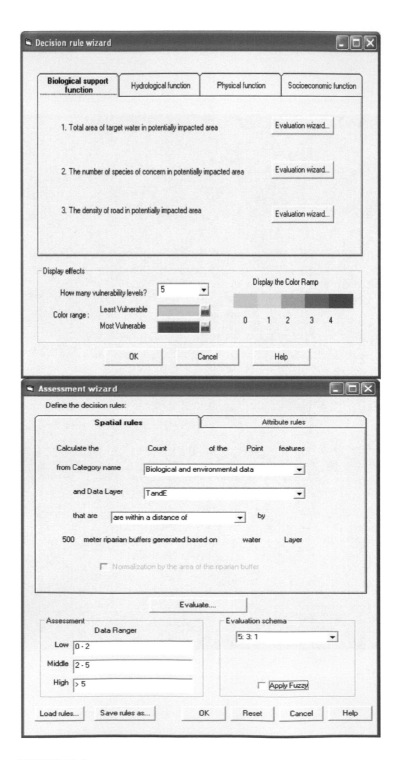

FIGURE 18.6

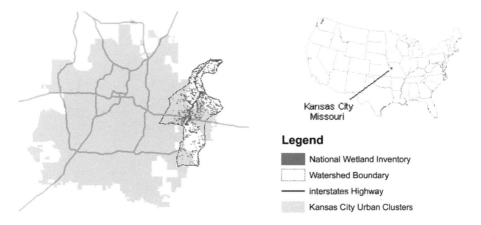

Legend

- National Wetland Inventory
- Watershed Boundary
- interstates Highway
- Kansas City Urban Clusters

FIGURE 18.7

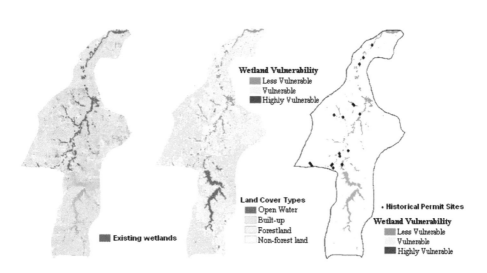

Existing wetlands

Wetland Vulnerability
- Less Vulnerable
- Vulnerable
- Highly Vulnerable

Land Cover Types
- Open Water
- Built-up
- Forestland
- Non-forest land

• Historical Permit Sites

Wetland Vulnerability
- Less Vulnerable
- Vulnerable
- Highly Vulnerable

FIGURE 18.8

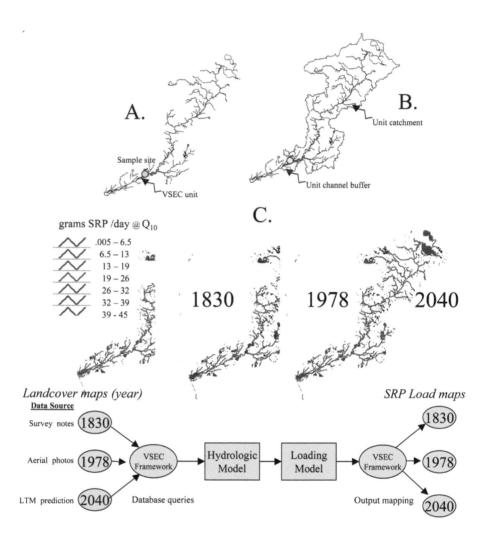

A.

Sample site

VSEC unit

B.

Unit catchment

Unit channel buffer

C.

grams SRP /day @ Q_{10}

.005 – 6.5
6.5 – 13
13 – 19
19 – 26
26 – 32
32 – 39
39 - 45

1830

1978

2040

Landcover maps (year)

Data Source

Survey notes (1830)

Aerial photos (1978)

LTM prediction (2040)

VSEC Framework

Database queries

Hydrologic Model

Loading Model

VSEC Framework

Output mapping

SRP Load maps

(1830)

(1978)

(2040)

FIGURE 20.4

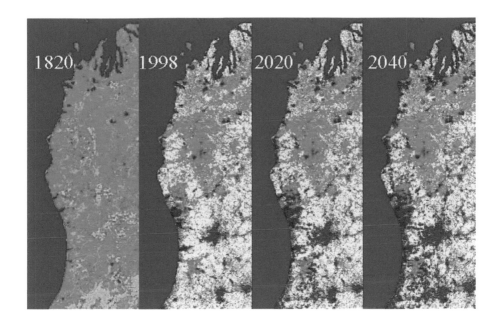

FIGURE 20.5

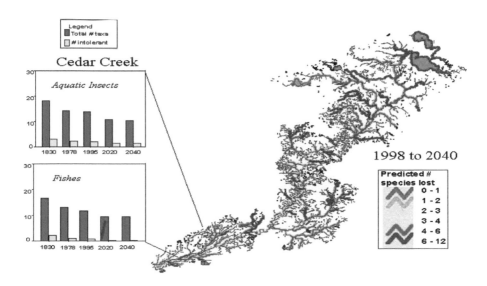

FIGURE 20.6

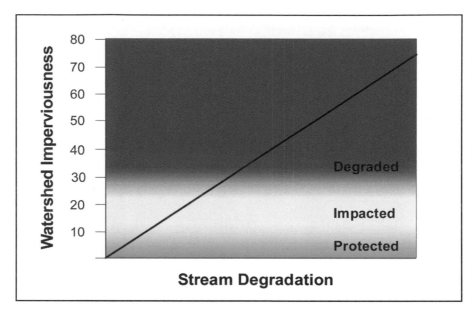

FIGURE 21.3

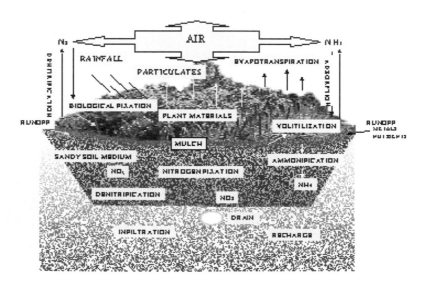

FIGURE 21.5

quantity of soil erosion in the wide area, but GIS provided a very useful environment to undertake the task of data compilation and analysis within a short period at any resolution.

The results suggest that soil erosion in Xiushui watershed is one of the major problems that hinder the full development of the local economy and should be more closely watched. The eroded particles and resultant sediments have already endangered the ecosystem of Poyang Lake wetland and more effective measures should be taken to change the situation.

ACKNOWLEDGMENTS

This work was supported by the National Key Basic Research and Development Program, 2003CB415205 and the Open Fund of Key Lab of Poyang Lake Ecological Environment and Resource Development (Jiangxi Normal University), Grant No. 200401006(1). The authors would like to thank Liu Ying for providing images and some other materials concerning Xiushui watershed.

REFERENCES

Angimaa, S. D., D. E. Stott, M. K. O'Neill, C. K. Ong, and G. A. Weesies. 2003. Soil erosion prediction using RUSLE for central Kenyan highland conditions. *Agriculture, Ecosystems and Environment* 97: 295–308.

Cai, C. F., S. W. Ding, Z. H. Shi, et al. 2000. Study of applying USLE and geographical information system IDRISI to predict soil erosion in small watershed. *Journal of Water and Soil Conservation* 14(2):19–24.

Desmet, P. J., and G. Govers. 1996. A GIS-procedure for automatically calculating the USLE LS-factor on topographically complex landscape units. *Journal of Soil and Water Conservation* 51(5):427–433.

Desmet, P. J., and G. Govers. 2000. *USLE2D.EXE (Release 4.1): User documentation*. Leuven: Catholic University of Leuven, Experimental Lab of Geomorphology.

Folly, A., M. C. Bronsveld, and M. Clavaux. 1996. A knowledge-based approach for C-factor mapping in Spain using Landsat TM and GIS. *International Journal of Remote Sensing* 12:2401–2415.

Fournier F, 1960. *Climat et érosion*. Paris: Ed. Presses Universitaires de France.

Liang, Y., and X. Z. Shi. 1999. Soil erodible K in East Hillyfields of the southern Yangtze River. *Research of Soil and Water Conservation* 6(2):47–52.

Lufafa, A., M. M. Tenywa, M. Iasbirye, M. J. G. Majaliwa, and P. L. Woomer. 2003. Prediction of soil erosion in a Lake Victoria basin catchment using a GIS-based universal soil loss model. *Agricultural Systems* 76:883–894.

Ma, Y. L., and L. H. Mei. Loss of water and erosion of soil in the Poyanghu Lake area and its prevention and cure measure. *Journal of Geological Hazards and Environment Preservation*, 2003, 14(3):31–35.

Mati, Bancy M., Royston P. C. Morgan, Francis N. Gichuki, John N. Quinton, Tim R. Brewer, and Hans P. Liniger. 2000. Assessment of erosion hazard with the USLE and GIS: A case study of the Upper Ewaso Ng'iro North basin of Kenya. *International Journal of Applied Earth Observation and Geoinformation* 2(2):78–86.

Meyer, L. D., A. Bauer, and R. D. Heil. 1985. Experimental approaches for quantifying the effect of soil erosion on productivity. In *Soil erosion and crop productivity*. ed. R. F. Follett, B. A. Stewart, and I. Y. Ballew. Madison, WI: American Society of Agronomy, Crop Science Society of America and Soil Science Society of America Publishers, 213–234.

Oyedele, J. D. 1996. Effects of erosion on the productivity of selected southwestern Nigerian soils. Ph.D. thesis, Department of Soil Science, Obafemi Awolowo University, Ile-Ife, Nigeria.

Tweddale, S. C., C. R. Echlschlaeger, and W. F. Seybold. 2000. An improved method for spatial extrapolation of vegetative cover estimates (USLE/RUSLE C factor) using LCTA and remotely sensed imagery. USAEC Report No. SFIM-AEC-EQ-TR-200011, ERDC/CERL TR-00-7. Champaign, IL: U.S. Army Engineer Research and Development Center, CERL.

USGS (U.S. Geological Survey). 2004. MODIS Terra vegetation indices. http://edcdaac.usgs.gov/modis/mod13q1.asp.

Wischmeier, W. H., and D. D. Smith. 1978. *Prediction rainfall erosion losses: A guide to conservation*, Agricultural Handbook 537. Washington, DC: Planning, Science and Education Administration, U.S. Department of Agriculture.

You, S. C., and W. Q. Li. 1999. Estimation of soil erosion supported by GIS: A case study in Guanji township, Taihe, Jiangxi. *Journal of Natural Resources* 14(1):62–68.

14 Evaluation of Rapid Assessment Techniques for Establishing Wetland Condition on a Watershed Scale

Vanessa L. Lougheed, Christian A. Parker, and R. Jan Stevenson

14.1 INTRODUCTION

Recently, the U.S. National Research Council (2001) recommended utilizing a watershed perspective together with science-based, rapid assessment procedures to track wetland mitigation and restoration. Rapid assessment tools can be used as a warning sign to give a quick idea of wetland condition and determine sites in need of further assessment or immediate protection. Many U.S. states have or are developing three-tiered assessment procedures that include an initial landscape-scale assessment using aerial imagery (tier 1), followed by a rapid condition assessment (tier 2), and a more intensive monitoring program (tier 3) (e.g., Miller and Gunsalus 1999, Mack 2001, Fennessy et al. 2004).

Wetlands can be significantly impacted by a variety of physical, chemical, and biological factors, and although a single environmental factor can sometimes be implicated as the primary stressor to a wetland ecosystem (King and Richardson 2003), it is more likely that a combination of factors result in wetland degradation on a landscape scale (Danielson 2001).

Furthermore, spatial and temporal variability in chemical stressor levels can make it difficult to diagnose one specific nutrient causing impairment, especially for sites sampled just once in landscape-scale assessments. In such cases, multistressor axes can be used to ensure assessments reflect a greater number of stressors (e.g., Mack 2001, Lougheed et al. 2001). In particular, one encounters a variety of wetland classes in a single watershed (e.g., lacustrine, riverine, and isolated wetlands) and these different classes may respond differently to a variety of stressors (Fennessy et al. 2004). Multistressor axes may therefore have a greater utility for a suite of wetlands in a landscape setting than does any one individual measure.

Existing rapid assessment methods generally combine various measures of hydrology, water quality, soils, landscape setting, and vegetation (Fennessy et al.

2004). Fennessy et al. (2004) reviewed 16 different rapid assessment methods that met 4 criteria they deemed to be important for successful rapid assessment. They concluded that the best methods should:

1. describe the condition along a single continuum ranging from least to most impacted
2. provide an accurate assessment of conditions in a relatively short time period (e.g., 1 day total for both field and lab components)
3. include an onsite assessment
4. be capable of onsite verification using more comprehensive ecological assessment data (tier 3)

Using these guidelines, the goal of this study was to develop a suite of rapid assessment techniques and examine their utility in evaluating wetland condition in a single large watershed in Michigan. In particular:

We compare a field-based estimate of riparian land use to actual land use values determined from GIS (geographic information system) maps in a 1-km buffer around each wetland.

We create a multimetric wetland disturbance axis (WDA) that incorporates rapid measures of hydrology, water quality, and land use.

As a rapid assessment of biological condition, we compare an estimate of epiphytic algal thickness against epiphytic chlorophyll biomass values and percent cover of epiphytic macroalgae.

To verify the utility of the WDA in reflecting biological condition, we determine whether plant community composition responds along the WDA.

14.2 METHODS

The Muskegon River drains a 7,000-km² watershed that flows into Lake Michigan on its eastern shore and is dominated by forested land in the upstream regions and agricultural and small urban areas (e.g., Muskegon, population 40,000) in the downstream region. We visited 85 wetlands in the Muskegon River watershed (MRW in Michigan) during the summers of 2001 through 2003. This included 35 isolated depressions, 25 lacustrine and 25 riverine wetlands. Fifty-two (52) sites were selected randomly based on a numbered grid overlaid on GIS-based wetland maps, while the remaining 33 sites were purposely selected to represent a gradient of disturbance. Approximately half (18) of the randomly selected wetlands were outside the MRW and in the upstream reaches of immediately adjacent watersheds (e.g., Chippewa River, Grand River, Pere Marquette River).

For determination of water chemistry, water was collected from an open water area in each wetland in 250-mL, acid-washed bottles. Total phosphorus (TP), total nitrogen (TN), nitrate + nitrite (NOx), ammonia (NH₃), silica (Si), soluble reactive phosphorus (SRP), and chloride (Cl) were determined using standard methods (American Public Health Association [APHA] 1998) on a Skalar auto-analyzer. Conductivity was measured in the field using a YSI 556 multiprobe. Sediment was collected from 3 random locations in the wetland using a 5-cm corer; the 3 samples were combined and frozen until analysis. C:N was determined using a Perkin-Elmer 2400 Series II CHN analyzer, while percent organic matter was determined following loss-on-ignition at 500°C. We did not measure contaminant levels in this study; however, local public health departments had identified several areas with contaminated sediments at a level of concern and these were noted.

We constructed a multimetric stressor axis designed to integrate and give equal weight to measurements in 3 primary stressor categories: land use, hydrological modification, and water quality. Unlike many other rapid assessment methods (see Fennessy et al. 2004), we did not include plant habitat variables, as we felt that this would create circular relationships with our plant community metrics. This wetland disturbance axis (WDA) included 3 metrics indicative of land use and land cover change (riparian land use, buffer width, distance to nearest wetland), 2 metrics indicative of hydrology (hydrological modification, water source), as well as 2 water quality metrics (conductivity, contaminants) (Table 14.1). Some of these metrics were loosely based on those used in the Ohio Rapid Assessment Method (ORAM) (Mack 2001), while new metrics were also included to reflect different data collection methods in this study. We assigned scores to some of the metrics by placing the "answers" to assessment questions into different categories and then assigning a score by category (Fennessy et al. 2004). For example, hydrological modification was categorized using questions such as: Are there roads along the wetland edge? Is there evidence of dams, dredging, or ditching? Then, each hydrological stressor answer was assigned a score, which was summed to achieve a metric indicative of all hydrological modifications. Most metrics were scaled using a 1-3-5 scaling system, where a value of 0 or 1 was given to the least impacted wetlands and a value of 5 was given to the most degraded sites. For example, average buffer width around wetlands was categorized in the field in 4 categories (0 = >50 m; 1 = 25–50 m; 3 = 10–25 m; 5 = <10 m). Similarly, water source was characterized as year round (0), intermittent (3), or none visible (5), and contaminants were classified as none (0), low levels (3), or level of concern (5). In the field, riparian land use was categorized as either agricultural, fallow pasture, urban, suburban, parkland, or forested on a scale from 0 to 4 (sum total of all categories = 4). For inclusion in the WDA, the proportion (out of 4) for each of these land use categories was multiplied by 5 (for high-impact land categories such as urban and agricultural land), by 3 (for moderate land use impacts such as fallow pasture, park, and suburban residential), whereas forested land was multiplied by zero. Two metrics (nearest neighbor, conductivity) were scaled based on the frequency distribution of values observed for all wetlands in this study. One of these, conductivity, was scaled from 0 to 10 in order to increase the weight of this metric in the overall WDA calculation. Finally, all individual scores from each metric were added together. Although the maximum WDA in this study was 75, the WDA was scaled from 0 to 100, to allow its use in more degraded watersheds in the region. Low value of the WDA indicate higher-quality wetlands.

Land use and distance between wetlands were determined in ESRI ArcMap (version 9.0) using land use maps current to 1998. Using these data, we determined linear distance to the nearest wetland (nearest neighbor), as well as riparian land use in a 1-km buffer around each wetland. Nearest neighbor is the only metric included in

TABLE 14.1

**Description of metrics used in the wetland disturbance axis (WDA).
Sum of all metrics is 45, but is scaled out of 100 to get final WDA.**

	Score and range of values	MAX
Land use and habitat fragmentation (MAX: 15)		
Average buffer width (score 1 value only)	0: >50 m 1: 25–50 m 3: 10–25 m 5: <10 m	5
Surrounding land use (calculate and add)	0: multiply 0x proportion forested land 3: multiply 3x sum of proportion park, fallow pasture, and suburban residential land 5: multiply 5x sum of proportion urban, industrial, and agricultural land	5
Nearest neighbor[a] (score 1 value only)	0: <0.13 km 1: 0.13–0.33 km 2: 0.33–0.66 km 3: 0.66–0.92 km 4: 0.92–1.64 km 5: >1.64 km	5
Hydrology (MAX: 15)		
Water source (score 1 value only)	0: year-round inputs (river, lake, groundwater) 3: seasonally intermittent 5: no visible inputs	5
Hydrological modification (add all visible modifications together to maximum of 10)	0: none 1: road along less than 1/4 of wetland edge 1: human dam (pre-1980) 3: human dams (post-1980) or natural dams (beaver, clogged culvert) 3: road along >1/4 of wetland edge 5: high impact (ditching, dredging, culverts)	10
Water quality (MAX: 15)		
Conductivity[a] (score 1 value only)	0: <85 µS/cm 2: 85–159 µS/cm 4: 159–289 µS/cm 6: 289–386 µS/cm 8: 386–498 µS/cm 10: >498 µS/cm	10
Contaminants (score 1 value only)	0: None 3: Present at low levels 5: Level of concern	5

[a] Ranges included in metric based on frequency distribution.

the WDA that was not estimated in the field; however, it may be possible to estimate this variable more rapidly using aerial photos or topographic maps if GIS is not available.

Macrophyte and epiphytic algae communities were surveyed using a stratified random design. We established 3 regularly spaced parallel transects, perpendicular to the shore, and randomly placed 1-m^2 rectangular quadrats along each transect according to a random numbers table. In each quadrat, we recorded relative cover of each plant species using a modified Braun-Blanquet scale, estimated the percent cover of filamentous macroalgae, and classified epiphyte thickness on a semiquantitative scale (rapid epiphyton survey [RES]: 0 = no growth; 1 = thin film, tracks can be drawn with your fingernail; 2 = 1 to 5 mm; 3 = >5 mm). These were visual estimations of epiphytic thickness, and did not represent precise measurements. Epiphytic algae were collected from cuttings of the dominant vegetation type in each wetland selected from random locations along each transect; we avoided collecting plants with macroalgal growth. Algae was removed from the plants with a combination of gentle rubbing from emergent stems and shaking of submerged plant stems in distilled water. Cleaned plants were placed in zipper bags and refrigerated so that surface area could be determined using image analysis software (ImageJ, NIH). Subsamples of the resulting algal suspension were frozen and analyzed for chlorophyll-a within 2 months of collection. Chlorophyll-a was extracted with 90% ethanol for 24 hours in the dark at 4°C; samples were then sonicated for 15 minutes and chlorophyll fluorescence determined on a Turner Designs fluorometer. Chlorophyll concentration was expressed per surface area of plant. Results presented are not corrected for phaeophytin because our RES could not distinguish between live and dead epiphytes.

We selected the Floristic Quality Assessment Index (FQAI) for Michigan (Herman et al. 2001) and its related coefficient of conservatism (CofC) to describe the wetland condition represented by the plant communities. The FQAI indicates the extent to which the community is dominated by sensitive wetland plants. The CofC is the sensitivity value given to each plant and we used the average CofC calculated for all plant species in each wetland. To explain structure in the biological communities of the wetlands, independent of any preconceived environmental preferences or gradients, we used nonmetric multidimensional scaling (NMDS). NMDS analysis identifies axes that describe biologically meaningful, multivariate gradients in the community data (McCune and Grace 2002). We selected the Bray-Curtis distance measure and used the first NMDS axis identified by PC-ORD (version 4.10) as an indicator of plant community structure. The NMDS, FQAI, and CofC were determined from previous analyses (Lougheed et al. 2007) to respond strongly to environmental gradients in the MRW.

Relationships between the rapid assessment variables and more detailed measurements of land use and epiphytic chlorophyll-a were studied in the large dataset of 85 wetlands, regardless of wetland class. In studying the responses of the plant communities, we divided the data into wetland classes (depressions, lacustrine, riverine) because biological communities in differing classes may respond uniquely to differing stressors.

14.3 RESULTS

Actual land use in a 1-km buffer around each wetland was well represented by the estimated land use categories (Figure 14.1); however, our estimates of land use more accurately reflected urban and agricultural land use. For both these land use types, we were able to distinguish among 3 separate categories and the 0 category had an average of 4% developed land in both cases. Our measurements of forested land differed between the lowest (0 and 1) and highest categories (3 and 4); however, the 0

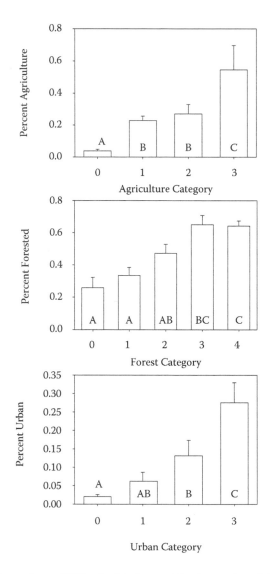

FIGURE 14.1 Comparison of GIS-calculated percent land use values determined in a 1-km buffer around each wetland in 4 to 5 land use categories estimated in the field. Letters indicate statistical similarities (Tukey multiple comparisons; p < 0.05).

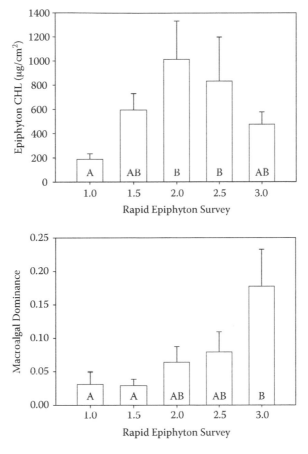

FIGURE 14.2 Comparison of epiphytic chlorophyll-a biomass (top) and macroalgal dominance (bottom) in 5 rapid epiphyton survey (RES) categories estimated in the field. Letters
indicate statistical similarities (Tukey multiple comparisons; p < 0.05).

category had an average of 26% forested land, which was not significantly different
from category 1 at 33% forested land.

We took the average rapid epiphyton survey (RES) values from each wetland
and rounded the value up to the nearest 0.5. Epiphytic chlorophyll-a was significantly different between sites with a "thin film" (category 1) of algae, relative to
sites with approximately 1 to 5 mm of growth (category 2) (Figure 14.2). There was
no significant increase in category 3, likely because it included sites with increased
macroalgal cover, which we excluded from our epiphyte samples. This is supported
by comparisons of macroalgal cover, expressed as relative dominance of macroalgal
cover (per m^2) relative to total plant species cover (per m^2), which was significantly
higher in sites with an average RES value of 3.

We used principal components analysis to determine which rapid assessment
metrics explained the greatest amount of variation in the dataset. The first 3 PCA
axes together accounted for 68% of the variation among sites. The first principal
component (PC1) explained 34% of the variation in the dataset, and was most highly

TABLE 14.2
Significant correlations between
WDA and environmental variables
(p < 0.10; Bonferoni corrected).

Variable	r	p
% Urban	0.54	0.0000
% Agriculture	0.43	0.0000
% Forest	−0.60	0.0000
TP	0.30	0.0061
NO_X	0.36	0.0008
SRP	0.21	0.0496
NH_3	0.35	0.0011
Cl	0.71	0.0000
Sediment: %organic	−0.23	0.0477
Sediment C:N	0.25	0.0356

correlated with land use and fragmentation variables (buffer width, r = 0.82; riparian land use, r = 0.83, nearest neighbor, r = 0.53) and water conductivity (r = 0.64); all other metrics were also significantly correlated with this axis, but at much lower levels (r = 0.24–0.36). The second axis (PC2; 18% of variation in dataset) was most highly correlated with hydrological variables (modification, r = 0.73; water source, r = 0.60), as well as a negative correlation with nearest neighbor (r = −0.59). The third axis (PC3; 16% of variation in dataset) was most highly correlated with contaminants (r = 0.80). There was no significant difference in the location of different wetland classes along PC1; however, PC2 values were significantly higher in depressional wetlands, likely because fewer of these had year-round inputs of water (water source) as opposed to all riverine and lacustrine sites.

As an indicator of disturbance, the WDA correlated strongly with many measured land use and water chemistry variables (p < 0.05). In particular, it was highly correlated with land use variables (r = 0.54–0.60), water chemistry measures (r = 0.3–0.36), and sediment characteristics (r = 0.23–0.25) (Table 14.2).

For subsequent analyses, we separated the data into 3 sections, representing the different wetland classes (depressions, lacustrine, riverine). A significant amount of the variation in a measure of plant community structure (NMDS) and the extent to which the community was dominated by native, sensitive taxa (FQAI and CofC) could be explained by the WDA (Figure 14.3). The FQAI was strongly correlated with the WDA for riverine sites, whereas the CofC was a better metric for depressions and lacustrine wetlands. Overall, these relationships were strongest for depressional and lacustrine wetlands, and lower for riverine sites. In many cases, the WDA explained more variation in the biological metrics than did any individual environmental variable (Table 14.3); however, forested land explained slightly more of the variation in the NMDS values for depressional wetlands, and variation in riverine plant communities was explained slightly better by TP and conductivity. It is interesting to note, however, that using a suite of 120 plant metrics calculated for all

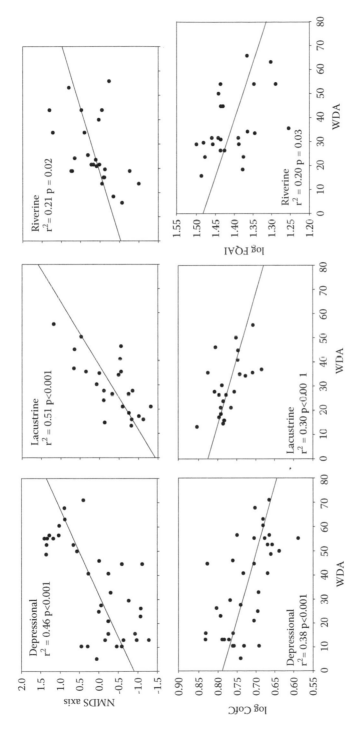

FIGURE 14.3 Relationship between plant community metrics and the WDA for depressional (left), lacustrine (middle), and riverine (right) wetlands in the MRW.

TABLE 14.3

Significant correlations between biological metrics & environmental variables (p < 0.10; Bonferoni corrected).

	Depressions		Lacustrine		Riverine	
	NMDS	CofC	NMS	CofC	NMS	FQAI
WDA	0.68	–0.61	0.77	–0.62[a]	0.50[a]	–0.55
Agriculture	0.61	–0.56	—	–0.54[a]	0.44[a]	—
Urban	0.39[a]	—	0.63	–0.41[a]	—	—
Forest	–0.72	0.55	–0.61	0.55	—	—
TP	0.36	—	0.501	–0.45[a]	0.54[a]	—
Cl	0.50	–0.39	0.62	–0.51[a]	—	–0.52[a]
COND	0.60	–0.46	0.65	—	0.41[a]	–0.63
NO$_x$	—	—	—	—	0.41[a]	–0.58

[a] Not significant when Bonferoni corrected at p < 0.05.

site types, including measures of species richness and plant community composition (Lougheed, unpublished data), more of these metrics were correlated to the WDA (26 metrics; Bonferoni corrected; p<0.05), than the next most commonly correlated environmental variables: developed land (21 metrics), TP (12 metrics), and Cl (5 metrics).

14.4 DISCUSSION

This study provides evidence that field-based estimates of algal cover and land use can accurately reflect more detailed measures requiring increased lab processing time and technical skills. In addition, we present the development and verification of a multimetric wetland disturbance axis (WDA) that successfully integrates stressors from 3 categories: land use, hydrological modification, and water quality. The WDA is highly correlated with a variety of land use and water chemistry measures, as well as several measures of plant community composition.

Rapid epiphyton assessment can be highly useful because it enables the determination of algal biomass over larger spatial scales than sampling algae off individual substrates followed by lab analysis (Stevenson and Bahls 1999). We provide evidence that an estimate of epiphyte cover using a rapid epiphyton survey can be a good surrogate for more detailed measures of epiphytic and macro-algal biomass. Despite its accuracy, both the rapid and more detailed measurements of algal biomass were not correlated to any rapid or detailed measures of wetland condition, including the WDA or nutrient levels. Wetlands are complex environments, where both vascular plants and algae compete for nutrients and light. Measures of diatom community composition (Lougheed et al. 2007) or trophic state indices (e.g., Van Dam et al. 1994) may be more sensitive indicators of algal responses to nutrient enrichment in wetlands than more simple measures of algal biomass. In particular, Lougheed et al. (2007) found that diatom community composition (as indicated by NMDS) was a highly sensitive measure of disturbance in depressional wetlands. Early changes in algal species composition, as opposed to changes in algal biomass, may result from minor changes in nutrient availability and may be a better indicator of alterations in fundamental microbial processes (Stevenson et al. 2002).

The proportion of agricultural and urban land in wetland watersheds is a highly significant predictor of reduced water quality in wetlands (Crosbie and Chow-Fraser 1999, Lougheed et al. 2001), while an increased proportion of forested land, including forested buffer strips along streams (e.g., Crosbie and Chow-Fraser 1999) in wetland watersheds, can be beneficial in improving water quality. Land use covers can be time-consuming to determine, especially if GIS layers are not available or experience using GIS programs is limited; however, this study indicates that riparian land use estimates can be a good approximation of actual riparian land use calculated from GIS layers. In this study, our estimates may have underestimated forested land in some cases, likely because many of our wetlands were accessible by roads or tracks and thus may have been biased toward sites closer to human habitations, even though overall land in the riparian area may have been largely forested. Nonetheless, these estimates appear to be a useful approximation of land use that may be used in riparian rapid assessment techniques.

A critical step in creating rapid assessment methods is ensuring their utility in reflecting wetland quality, not only at the level of the chemical and physical variables included in the method, but also of the more intensive biological monitoring that might occur in a tier 3 assessment. The WDA proved to be useful in integrating the effects of land use, hydrological alteration, and nutrient-based stressors in the MRW. In particular, much of the variation among sites in the MRW was due to differences in land use (including buffer width, riparian land use and nearest neighbor) and water conductivity. As verification of its utility, the WDA was highly correlated with detailed land use and water quality measures, as well as measures of plant community composition (NMDS) and dominance by sensitive plants (CofC, FQAI). In addition, Lougheed et al. (2007) showed that the WDA could be used as a rapid assessment tool for categorizing depressional wetlands into tiers to track restoration and degradation. They used nonlinear biological responses along the WDA to identify biological thresholds, and thus classified wetlands into 3 groups: reference sites with little biological change (WDA < 17), slightly altered sites (17 < WDA < 47) where the most sensitive organisms responded, and degraded sites (WDA > 47) where large-scale changes in community structure of plants, diatoms, and zooplankton occurred. Given these analyses for depressional wetlands in the MRW and nearby watersheds, the WDA rapid assessment tool, which has been verified using more comprehensive biological data, can now be used to categorize additional wetlands in the watershed as well as track the state of wetlands that were identified as needing remedial action.

In an era of reduced funding for environmental monitoring and research, combined with an increased need for monitoring the ever-increasing impacts of human activities, development and validation of rapid assessment techniques is necessary to allow for the assessment, protection, and restoration of aquatic habitats. The WDA meets all criteria necessary for successful rapid assessment of wetland sites including: (1) representing a continuum from least to most degraded, (2) it can be completed in a relatively short period of time using both onsite and lab components, and (3) it can be verified using comprehensive ecological assessment data (Fennessy et al. 2004). The WDA will be useful in tracking wetland quality in the MRW, and providing a warning sign to identify sites in need of immediate protection. In addition, we have provided evidence that field-based estimates of algal cover and land use can accurately reflect more detailed measures requiring increased lab processing time and technical skills.

ACKNOWLEDGMENTS

We greatly appreciate field assistance from Mollie McIntosh, Sarah Wolf, Alyson Yagiela, Nicole Behnke, and James Montante. Land use shapefiles were provided by Brian Pijanowski. This project was funded by the Great Lakes Fisheries Trust as part of the Muskegon River Initiative.

REFERENCES

American Public Health Association (APHA). 1998. *Standard methods for the examination of water and wastewater.* 20th ed. Washington, DC: American Public Health Association.

Crosbie, B., and P. Chow-Fraser. 1999. Percentage land use in the watershed determines the water and sediment quality of 22 marshes in the Great Lakes basin. *Canadian Journal of Fisheries and Aquatic Science* 56:1781–1791.

Danielson, T. J. 2001. *Methods for evaluating wetland condition: Introduction to wetland biological assessment.* EPA 822-R-01-007a. Washington, DC: Office of Water, U.S. Environmental Protection Agency.

Fennessy, M. S., A. D. Jacobs, and M. E. Kentula. 2004. *Review of rapid methods for assessing wetland condition.* EPA/620/R-04/009. Washington, DC: U.S. Environmental Protection Agency.

Herman, K. D., L. A. Masters, M. R. Penskar, A. A. Reznicek, G. S. Wilhelm, W. W. Brodovich, and K. P. Gardiner. 2001. *Floristic quality assessment with wetland categories and examples of computer applications for the state of Michigan*, revised 2nd ed. Lansing: Michigan Department of Natural Resources, Wildlife, Natural Heritage Program.

King, R. S., and C. J. Richardson. 2003. Integrating bioassessment and ecological risk assessment: An approach to developing numerical water-quality criteria. *Environmental Management* 31(6):795–809.

Lougheed, V. L., C. A. Parker, and R. J. Stevenson. 2007. Using non-linear responses of multiple taxonomic groups to establish criteria protective of wetland biological condition. *Wetlands* 27:96–109.

Lougheed, V. L., B. Crosbie, and P. Chow-Fraser. 2001. Primary determinants of macrophyte community structure in 62 marshes across the Great Lakes basin: Latitude, land use, and water quality effects. *Canadian Journal of Fisheries and Aquatic Sciences* 58:1603–1612.

Mack, J. J. 2001. *Ohio rapid assessment method for wetlands, manual for using version 5.0.* Ohio EPA Technical Bulletin Wetland/2001-1-1. Columbus: Ohio Environmental Protection Agency, Division of Surface Water, 401 Wetland Ecology Unit.

McCune, B., and J. B. Grace. 2002. *Analysis of ecological communities.* Gleneden Beach, OR: MjM Software Designs.

Miller, R. E., Jr., and B. E. Gunsalus. 1999. *Wetland rapid assessment procedure.* Updated 2nd ed. Technical Publication REG-001. West Palm Beach, FL: Natural Resource Management Division, Regulation Department, South Florida Water Management District, http://www.sfwmd.gov/newsr/3_publications.html.

National Research Council (NRC). 2001. *Compensating for wetland losses under the Clean Water Act.* Washington, DC: National Academy Press.

Stevenson, R. J., and L. L. Bahls. 1999. Periphyton protocols. In *Rapid bioassessment protocols for use in streams and wadeable rivers: Periphyton, benthic macroinvertebrates and fish,* 2nd ed., EPA 841-B-99-002, ed. M. T. Barbour et al. Washington, DC: U.S. Environmental Protection Agency; Office of Water.

Stevenson, R. J., P. V. McCormick, and R. Frydenborg. 2002. Methods for evaluating wetland condition: Using algae to assess environmental conditions in wetlands, EPA-822-R-02-021. Washington, DC: Office of Water, U.S. Environmental Protection Agency.

Van Dam, H., A. Mertenes, and J. Sinkeldam. 1994. A coded checklist and ecological indicator values of freshwater diatoms from the Netherlands. *Netherlands Journal of Aquatic Ecology* 28:117–133.

15 Development of Geospatial Ecological Indicators in Jiangxi Province, China

Peng Guo and Xiaoling Chen

15.1 INTRODUCTION

Comprehensive evaluation of ecological environments is necessary for environmental sustainability and management planning, which can provide quantitative documents as scientific guides for informed decision making. The existence of several environmental impact assessment (EIA) methods makes it difficult to make on an appropriate choice (Sankoh 1996).

Remote sensing, being a very useful observational tool, has been integrated with GIS (geographic information system) to monitor and evaluate environmental conditions (Shen et al. 2004, Wang et al. 2002, 2004). In these studies, Landsat TM, ETM+ and NOAA/AVHRR (National Oceanic and Atmospheric Administration/Advanced Very High Resolution Radiometer) images were used to obtain the vegetation cover status derived from NDVI (Normalized Distance Vegetative Index). In this paper, we demonstrate the potential of the Moderate Resolution Imaging Spectroradiometer (MODIS) for monitoring ecological and environmental conditions. MODIS is flown on two NASA satellites (Terra/morning pass: 1999–present and Aqua/afternoon pass: 2002–present). MODIS data are collected continuously and are available for public use.

Poyang Lake, located in the middle reach of the Yangtze River and the subtropical wet monsoon zone, is the largest freshwater lake in China. The area of Poyang Lake basin is $16.22 \times 10^4 \, km^2$ (9% of the Yangtze River basin), and 97% of the basin is located in Jiangxi Province. Over many years, the total area of forest and lake land cover has been reduced due to human activities such as increasing population, which has resulted in the deterioration of the ecological environment in the Poyang Lake basin. The area of soil and water being damaged by erosion in Jiangxi Province is $3.52 \times 10^4 \, km^2$ (Guan 2001). This process of erosion has been seriously destroying the quality of the ecological environment and impacting the sustainable development of economic growth in Jiangxi Province and the Yangtze River basin. This paper focuses on developing geospatial ecological indicators and quantitatively analyzing environmental factors of Jiangxi Province based on remote sensing and GIS methods.

15.2 MATERIALS AND METHODS

15.2.1 BUILDING AN EVALUATION INDICATOR SYSTEM

Many factors, including biophysical and anthropogenic factors, may influence the ecology and environment in the basin. In order to quantitatively depict the characteristics of ecology and environment, an evaluation indicator system needs to be developed in a comprehensive and concise way for factor selection. In this chapter, various factors, such as characteristics and spatial scale of the Jiangxi Province, climate, terrain, soil erosion, and vegetation cover status, were selected to build the indicator system that can better reflect the ecological and environmental characteristics of the study area. The data mainly includes $\geq 0°$ accumulated temperature, $\geq 10°$ accumulated temperature, average annual temperature, precipitation and evapotranspiration, surface humid index, vegetation index, elevation, slope, aspect, soil erosion, and MODIS image.

15.2.2 EXTRACTION OF WATER BODIES AND NDVI

In the process of evaluation, a water body was regarded as an individual land object to be distinguished from others. According to the characteristics of spectral reflectance of a water body and vegetation, the near-infrared band is the most useful in distinguishing the land-and-water boundary and ground vegetation (Zhen and Chen 1995, Tan et al. 2004). Bands 1 and 2 of the MODIS images, whose wavelengths range from 0.62 to 0.67 μm and 0.84 to 0.87 μm, respectively, match bands 3 and 4 of the Landsat TM images. These two bands were used to identify vegetation and water body. The NDVI is adopted to synthesize vegetation index diagram as follows:

$$NDVI = (NIR - Red)/(NIR + Red) = (Band2 - Band1)/ (Band2 + Band1)$$

The *NIR* and *Red* are are digital numbers (DNs) of the near-infrared band and red band, respectively. The analysis showed that the area of NDVI <0.25 was covered by water bodies in the wetlands of the Poyang Lake basin.

15.2.3 STATISTICAL METHODS

It is important and difficult to integrate the multi-indices into an integrative indicator when the ecological environment is evaluated. Some methods such as weighted index, AHP (analytical hierarchy process), and GEM (group eigenvalue method) have been ordinarily used. In this chapter, PCA (principal component analysis) was performed on the ecological parameters to evaluate the ecological environment in Jiangxi Province.

15.2.4 PRINCIPAL COMPONENT ANALYSIS

Principal component analysis is a powerful technique for pattern recognition that attempts to explain the variance of a large set of intercorrelated variables and transform them into a smaller set of independent (uncorrelated) variables (principal components). The first principal component accounts for as much of the variability in the dataset as possible, and each succeeding component accounts for as much of the remaining variability as possible. Principal component analysis provides information on the most meaningful parameters, which describe whole datasets, then render data reduction with minimum loss of original information.

Performed with GIS software, spatial PCA is used to transform the data in a stack from the input multivariate attribute space into a new multivariate attribute space whose axes are rotated with respect to the original space. The axes (attributes) in the new space are uncorrelated. The main reasons for transforming the data in a principal component analysis are to "compress" data by eliminating redundancy, to emphasize the variance within the grids of a stack, and to make the data more interpretable.

In the evaluation of an ecological environment, the integrative indicator can be defined as a weighted sum of M principal components. The weight is the component loading of principal components. The function is expressed as:

$$E = a_1Y_1 + a_2Y_2 + \cdots + a_MY_M \tag{15.1}$$

where E is the result of assessment, a is the component loading, and Y is the principal component. For excluding the effect of different scalars and dimensions of variables, the data were standardized by the follow equation:

$$D_i = \frac{(X_i - X_{aver})}{\sigma} \tag{15.2}$$

where X_{aver} is the average of X, and σ is the standard deviation of X.

15.2.5 QUALITY INDEX OF ECOLOGICAL ENVIRONMENTAL BACKGROUND

Based on the result of an ecological environmental synthetic evaluation, we used the following function to calculate the quality index of the ecological environment background of Jiangxi Province:

$$I_j = 100 \times \sum_{i=1}^{6} E_i \times (A_i / S_i) \tag{15.3}$$

where I_j is the quality index of the ecological environment background in the region j; E_i is the value of eco-environmental classes, as there are six environmental classes; the eco-environment with the best situation is assigned as 6, and the one with the worst situation is assigned as 1; A_i is the area of class i and S_i is the area of region j.

15.3 RESULTS AND DISCUSSION

By processing data with GRID model in the ArcInfo workstation, we calculated the climate index using the following parameters: ≥0° accumulated temperature, ≥10°

accumulated temperature, the average annual temperature, precipitation and evapo-transpiration, surface humid index, and terrain index from the following parameters: elevation, slope, aspect.

The equations are as follows:

$$I_{climate} = 0.5149 \times ClimateP1 + 0.3040 \times ClimateP2 + 0.1721 \times ClimateP3 \quad (15.4)$$

$$I_{terr} = 0.6385 \times TerrP1 + 0.2582 \times TerrP2 \quad (15.5)$$

where $I_{climate}$ is the climate index, $ClimateP_i$ is the *ith* principal component of climate index; I_{terr} is the topographical index, and $TerrP_i$ is the *ith* principal component of topographical index.

After analyzing the vegetation index, soil erosion index, topographical index, and climate index by using the PCA method, we calculated the ecological environmental assessment index in Jiangxi Province. Table 15.1 shows the principal component Eigen values of various principal components.

From the above analysis, we derived the linear equation of the ecological environmental assessment index:

$$I_{eco} = 0.7806 \times EcoP_1 + 0.1407 \times EcoP_2 \quad (15.6)$$

where I_{eco} is the ecological environmental assessment index, $EcoP_1$ is the first PC (principal component) of four indices, and $EcoP_2$ is the second PC.

We then calculated the integrated ecological environmental evaluation for Jiangxi Province (Table 15.2) using the environmental evaluation classification map (Figure 15.1).

The ecological environment was classified into six classes by the eco-environment indicator. Table 15.2 shows that the largest area occurs in class 5, while the sum of classes 3, 4, and 5 covers 64.05% of the total area. The sixth class, describing the optimal situation of a quality eco-environment, covers 16.70% of the total province, which is a class characterized by less soil erosion and higher vegetation coverage.

Figure 15.1 shows that the quality of the ecological environment in the upper reaches of the five rivers in the Poyang Lake basin is worse than the lower reaches of those rivers in the basin, which are near Poyang Lake and mainly consist of flood-plains. The environment in the upper reaches of the five rivers in the Poyang Lake basin has a great potential for affecting the water quality of Poyang Lake. Figure 15.2 is the quality index map of the eco-environmental background of Jiangxi Province, which was calculated based on equation 15.3, which can be used to spatially identify the environmental quality in Jiangxi Province. The results revealed that the Nanchang area has good eco-environmental quality. In addition, we determined that Ganzhou in the southern part of the Poyang Lake basin, and Pingxiang in the northern part, have poor eco-environmental quality.

TABLE 15.1
The eigenvalues and attribute ratio in the principal components analysis.

PC	Ecological environment index			Terrain index			Climate index		
	Eigen values	Deviation loading (%)	Accumulated deviation (%)	Eigen values	Deviation loading (%)	Accumulated deviation (%)	Eigen values	Deviation loading (%)	Accumulated deviation (%)
I	1.37968	78.06	78.06	1.02512	63.85	63.85	1.42617	0.5149	51.49
II	0.24869	14.07	92.13	0.43053	25.82	89.67	0.59126	0.3040	81.89
III	0.12187	6.90	99.03	0.15990	10.33	100	0.49594	0.1721	99.10

TABLE 15.2

The integrated ecological environmental evaluation of Jiangxi Province, China.

Eco-environmental class	Grid number	Area (km²)	Percent (%)
1	775080	7750.8	4.64
2	2152866	21528.66	12.89
3	3199463	31994.63	19.16
4	3684927	36849.27	22.07
5	3810588	38105.88	22.82
6	2788534	27885.34	16.70
Water	284414	2844.14	1.70
Total	16765872	167658.72	100.00

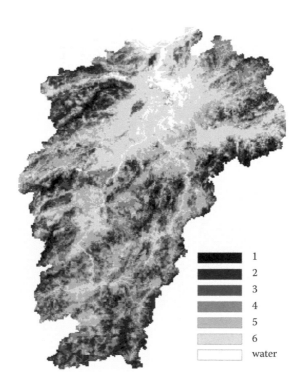

FIGURE 15.1 The integrated eco-environmental classes of Jiangxi Province, China. (*Source: Chinese State Bureau of Surveying and Mapping*)

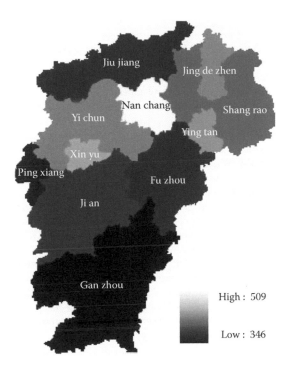

FIGURE 15.2   The quality index map of the eco-environmental background of Jiangxi Province, China. (*Source: Chinese State Bureau of Surveying and Mapping*)

15.4 CONCLUSIONS

It is obvious that remote sensing and geographic information science (RSGIS) are useful tools for studying the status of eco-environmental quality for diverse regions. RSGIS has many advantages such as lower cost, faster information collection, and more efficient investigation and assessment of the eco-environment on a large spatial scale. With the development of a new generation of remote sensing sensors, RSGIS will provide us with more feasible and cheaper means for studying the environmental problem. PCA, as a very useful analysis method, can be commendably used in the assessment. Combining the PCA method with RSGIS, this study performs primary research on the eco-environment in Jiangxi Province, which produced meaningful results. In next stage, we will study the change of eco-environment over the past 30 years by using this study frame.

ACKNOWLEDGMENTS

This work was supported by the National Key Basic Research and Development Program: 2003CB415205 and the Opening Foundation of the Key Lab of Poyang Lake Ecological Environment and Resource Development at Jiangxi Normal University, Grant No.200401006(1).

REFERENCES

Bernard, P., and L. Antoine. 2004. Principal component analysis: An appropriate tool for water quality evaluation and management-application to a tropical lake system. *Ecological modeling* 295–311.

Guan, R. S., 2001. Countermeasures on prevention and control of soil and water loss and effects towards flood control of Jiangxi Province. *Soil and Water Conservation in China* (10):21–22.

Sankoh, O. A. 1996. An evaluation of the analysis of ecological risks method in environmental impact assessment. *Environmental Impact Assessment Review* 16:183–188.

Shen, W., J. Zhang, and W. Wang. 2004. Ecological environmental quality assessment of the Three Gorges reservoir area based on remote sensing and GIS. *Resources and environment in the Yangtze Basin* 13(2):159–162.

Tan, Q. L., S. W. Bi, J. P. Hu, and Z. J. Liu. 2004. Measuring lake water level using multisource remote sensing images combined with hydrological statistical data. *Geoscience and Remote Sensing Symposium, 2004*, Vol. 7. Anchorage, AK: Institute of Electrical and Electronics Engineers, 4885–4888.

Wang, S. Y., Z. X. Zhang, and X. L. Zhao. 2002. Eco-environment synthetic analysis based on RS and GIS technology in Hubei province. *Advances in Earth Science* 17(3):426–431.

Wang, S. Y., G. Q. Wang, and Z. X. Chen. 2004. Eco-environmental evaluation and changes in Yellow River basin. *Journal of Mountain Science* 22(2):133–139.

Zhen, W., and S. P. Chen. 1995. *Introduction to resource remote sensing* [in Chinese]. Beijing: Press of Chinese Science and Technology, 103–168.

16 A Conceptual Framework for Integrating a Simulation Model of Plant Biomass for *Vallisneria spiralis* L. with Remote Sensing and a Geographical Information System

Guofeng Wu, Jan de Leeuw, Elly P. H. Best,
Jeb Barzen, Valentijn Venus, James Burnham,
Yaolin Liu, and Weitao Ji

16.1 INTRODUCTION

Submerged aquatic vegetation (SAV) forms a significant component of shallow lake and river ecosystems. This cycles nutrients, stabilizes flow, enhances water clarity, influences primary production, and protects the shores from erosion (Carr et al. 1997, Van Nes et al. 2003). Moreover, it provides an important food source for migratory birds. Bewick's swans (*Cygnus columbianus bewickii*), for instance, feed on the tubers of *Potamogeton pectinatus* in autumn in the Netherlands (Jonzen et al. 2002), while in Canada Trumpeter swans (*Cygnus buccinator*) forage on the tubers of *Potamogeton pectinatus* during spring (LaMontagne et al. 2003). Migratory North American waterfowl species in the tribe Athyini depend heavily upon tubers produced by a variety of SAV species (Korschgen 1989), and Siberian cranes (*Grus leucogeranus*) rely on the tubers of *Vallisneria spiralis* L. throughout the winter at Poyang Lake, China (Wu and Ji 2002). Careful management of lake and river systems with SAV are therefore indispensable not only from the perspective of ecosystem health, but also for the conservation of migratory bird species. Protected lake areas and backwater areas of rivers are not isolated water bodies, but are parts of larger watersheds. Natural or anthropogenic changes may influence these water bodies and, in turn, impact the growing conditions of SAV along with the availability of food for waterfowl. These changes can often be subtle. Van Vierssen et al. (1994), for example, described that SAV declined during the 1960s and 1970s in Lake Veluwe, the Netherlands, due to eutrophication caused by wastewater, and

FIGURE 16.1   Landsat TM images of Poyang Lake. In the dry season the water level is relatively low and numerous shallow lakes are disconnected from the main water body of Poyang Lake (left image captured on 28 October 2004), while in summer all these lakes compose one large water body (right image captured on 15 July 1989). **(See color insert after p. 162.)**

recovered following significant water quality improvement since the late 1980s. Harwell and Havens (2003) mentioned that elevated water levels and low water transparency during the late 1970s caused a decline in SAV density in Lake Okeechobee in southern Florida. Similar changes in SAV abundance have also occurred in the upper Mississippi River over fifty years (Jahn and Anderson 1986).

Poyang Lake (Figure 16.1), China, hosts an extremely high density of wintering birds, many of whom feed on *Vallisneria spiralis L.* tubers (Wu and Ji 2002). Large numbers of swans and virtually the entire world's population of Siberian cranes rely exclusively on this food source (Meine and Archibald 1996). *Vallisneria* tubers are also an important part of the diet of the endangered Swan Goose (Wu and Ji 2002). The production and availability of *Vallisneria* tubers depend on the local prevailing environmental conditions in both summer and winter. The environmental conditions, in turn, are not only influenced by the vegetation, birds, and activities of the local human community inside the protected areas, but also by the surrounding hydrological conditions. The Poyang Lake ecosystem is full of feedback systems operating among various biotic and abiotic factors (Figure 16.2). The hydrological conditions might also be changed due to ongoing and proposed engineering projects in the catchments of both Poyang Lake and the Yangtze River, as well as dredging as reported by Wu et al. (2007b).

Poyang Lake National Nature Reserve (PLNNR), designated with the highest conservation category in China, is the oldest protected area in Poyang Lake. It was established in 1988 to conserve the Siberian cranes, and places 5% of the total area of Poyang Lake under protected status. Proper management of this nature reserve requires the ability to forecast the impact of engineering projects on the productivity of *Vallisneria*. More importantly, the use of PLNNR by tuber-feeding birds varies greatly among, as well as within, years (Kanai et al. 2002). Given the spatial scale of the entire Poyang Lake, models would be the only possible tool that could be used for predicting future impacts of multiple development projects over such a large

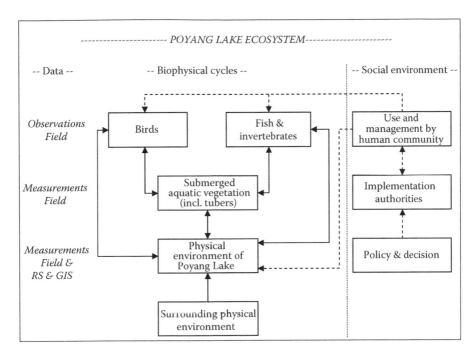

FIGURE 16.2 Hypothesized and important cause-and-effect interactions among the environment, submerged aquatic vegetation, and tuber-feeding birds in the Poyang Lake ecosystem, China. Solid lines are material flow; interrupted lines are activities and information flows; RS = remote sensing; GIS = geographic information system.

and heterogeneous landscape. In contrast, models required to develop this management tool are currently lacking in this region. Process-based shallow lake ecosystem models that incorporate hydrological variables and simulate SAV biomass would be ideally suited for such a purpose.

PLNNR is composed of nine lakes encompassing approximately 224 km² (Wu and Ji 2002), while the entire Poyang Lake basin encompasses 4,500 km². It would be costly and difficult to measure the onsite environmental conditions in every lake within the reserve, much less among the various regions of the entire lake that cranes use (Kanai et al. 2002). Remote sensing (RS) techniques and geographic information systems (GIS) might provide the means to quantify some of the variables required to drive the above-mentioned SAV biomass simulation models. In addition, a GIS could be used as a tool for the input, storage, retrieval, manipulation, analysis, and output of these spatial data, and also serve as an intermediary for implementing the SAV biomass simulation models.

The objective of this paper is to propose a conceptual framework for the integration of a SAV biomass simulation model for *Vallisneria* with RS techniques and a GIS. After implementation, the system could be used to study the response of the growth of *Vallisneria* to changes in environmental conditions, and predict the density and mass of *Vallisneria* tubers in space and time. The potential roles of GIS and RS in this endeavor will be emphasized. The questions to be addressed and potential user communities will be outlined and discussed.

16.2 ECOLOGY AND PRODUCTION OF *VALLISNERIA*

The production ecology of *Vallisneria* in Poyang Lake has been studied by Wu and Ji (2002). *Vallisneria* winters as tubers and seeds. It sprouts in April, flowers in June, reaches its maximum aboveground biomass in mid-September, and fills its tubers from mid-July until shoots senesce in late October. Field measurements in 1999 indicated that aboveground plant biomass decreased from the shore to the center of the lakes. The average shoot density close to the shore was about 50 to 70 stems/m², and individual fresh weight of each green shoot at maximum extent averaged about 12 g/stem. In areas with high tuber abundance, tuber density reached 10 to 70 tubers/m² with a biomass range of 6 to 45 g fresh weight/m². Wu and Ji (2002) concluded that water level, light intensity, temperature, and chemical properties of the water influenced growth and production of *Vallisneria* most. The importance of water level was demonstrated by the observation that the *Vallisneria* population collapsed during the flood of 1998, when water levels were more than four meters higher than in normal years (Wu and Ji 2002).

16.3 SAV SIMULATION MODELS

Relationships between environmental variables and the biomass of SAV have been included in a variety of plant growth simulation models (Collins and Wlosinski 1989, Scheffer et al. 1993, Calado and Duarte 2000, Best and Boyd 2001b, Van Nes et al. 2003, Herb and Stefan 2003, Giusti and Marsili-Libelli 2005). Some of these models are generic, such as Charisma (Van Nes et al. 2003), while others focus on one species, for example MEGAPLANT for *Potamogeton pectinatus L.* (Hootsmans 1994, Scheffer et al. 1993) and VALLA for *Vallisneria americana Michx* (Best and Boyd 2001a).

All these models include descriptions of the seasonal growth cycle. The modeled plants may survive the winter as shoots or as wintering structures (tubers or seeds). Growth is initiated at a preset day, and from that moment onward, each wintering structure transforms a fixed daily percent of its remaining biomass into the sprout. At a later preset day, aboveground biomass is transformed into belowground structures that survive the winter.

Environmental factors and vegetation characteristics influence growth. The macrophyte characteristics were modeled with different degrees of complexity. For example, Best and Boyd (2001a) considered different vertical layers within the shoot biomass, plant organs, and >1 tuber cohort sprouting per year, while Hootsmans (1994) took the formation of a second shoot into account.

Of particular relevance for our current case is VALLA (Best and Boyd 2001a), a model that has been developed for *Vallisneria americana*, a species closely related to *Vallisneria spiralis*, and can be run at climates varying from temperate to tropical. This model has been used to evaluate the feasibility for *Vallisneria americana* to recolonize rehabilitated navigation pools of the Illinois River (Best et al. 2004). It has also been used for ecological risk assessment of planned changes in commercial navigation on the upper Mississippi River (UMR) (Bartell et al. 2000). After recalibration for fast-flowing river conditions (Best et al. 2005), the model is currently being used to evaluate the tentative effects of changes in water level management in navigation pools of the UMR, USA, on the tuber availability for wintering water fowl (including Tundra swans). The latter UMR case is similar to the current Poyang Lake case, in which a major goal is to evaluate the tuber availability for the Siberian cranes and other tuber-grazing waterfowl.

Field data published by Wu and Ji (2002) suggest that the species of *Vallisneria* occurring in Poyang Lake is unclear. Wu and Ji (2002) also described the development cycle and showed the prevalent Poyang Lake *Vallisneria* species produces relatively small shoots and large tubers compared to the default *Vallisneria americana* species used for the latest recalibration of VALLA for the UMR. A clearer description of the Poyang Lake *Vallisneria* has been provided now, and new seeds are currently available, so the identification of the Poyang Lake *Vallisneria* species should be possible. Once VALLA is recalibrated for the Poyang Lake *Vallisneria* species, various scenarios of water level and transparency changes expected from development projects could be modeled and the impacts on the tuber production in Poyang Lake might be predicted. Other research on use of tubers by birds could then be incorporated into this model to predict changes to the carrying capacity of wintering waterfowl that feed on tubers.

16.4 CONCEPTUAL FRAMEWORK INTEGRATING THE SAV SIMULATION MODEL WITH RS AND GIS

Typically models have been used to simulate the impact of changing environmental conditions on SAV, and it would be tempting to combine these growth simulation models with GIS to make predictions across spatially heterogeneous environments. Huang and Jiang (2002) distinguished three approaches to integrating environmental models with GIS: loose, tight, and full coupling. Loose coupling relies on the transfer of data files between stand-alone GIS and environmental models. For tight coupling, the environmental models are usually embedded within a GIS, and the interactions between them depend on the parameter transfers. Full coupling is a full integration of environmental models within GIS, generally using an advanced programming language or GIS macro language.

We propose a conceptual framework (Figure 16.3) of loosely integrating VALLA with RS and a GIS for simulating the plant and tuber biomass of *Vallisneria*. This framework consists of three main components: GIS, RS, and the SAV simulation model. RS and GIS will be used to estimate spatial distributions of important environmental variables that vary frequently and have not been recorded in the field, such as solar radiation reaching the water surface and water temperature. Additionally, the GIS is also used to input, store, retrieve, manipulate, analyze, and output spatial variables and other spatial data, such as solar radiation, digital elevation model (DEM), light attenuation, and spatial distribution of tuber biomass. The SAV simulation model imports the environmental variables from GIS, simulates the plant and tuber production of *Vallisneria,* and exports the results to GIS for analysis and visualization.

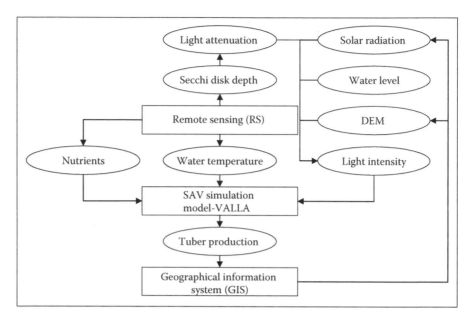

FIGURE 16.3   A conceptual framework integrating a SAV simulation model of *Vallisneria* with RS and a GIS. Solar radiation represents the global solar radiation reaching the water surface, and light intensity represents the light intensity reaching the canopy of SAV.

16.5 ENVIRONMENTAL VARIABLES AND THEIR MEASUREMENTS IN POYANG LAKE

SAV simulation models require input data such as light intensity reaching the canopy of the vegetation, water temperature, and the concentrations of nutrients and CO_2. The spatial distribution of none of these variables has been measured in Poyang Lake. However, a number of associated variables that could be used to derive light intensity reaching the canopy, such as water level and Secchi disk depth (SDD), have been measured. These in situ measurements, however, have been collected for a few lakes only. Confronted with such problems, the geoinformation technologies could be correlated with limited in situ sampling and be used to measure or infer environmental variables for large areas without in situ sampling.

The light intensity reaching the canopy of the *Vallisneria* is probably the most important variable to be measured since the availability of light is the primary factor controlling photosynthesis of SAV, a primary plant function that drives biomass growth in most aquatic systems (Carr et al. 1997). Light at the canopy level is determined by the radiation reaching the water surface, reflection by the water surface, and penetration of the light through the water column to the canopy of SAV (Figure 16.4). Presently no reliable solar radiation estimates exist for Poyang Lake, and the closest station with such data, Nanchang, is 65 kilometers away.

Measurements of sunshine hours (S_{+0}) are available for eight stations around Poyang Lake. Using these data we combined, with a triangulated irregular network (TIN) and inverse distance weighting (IDW) interpolation method, interpolated

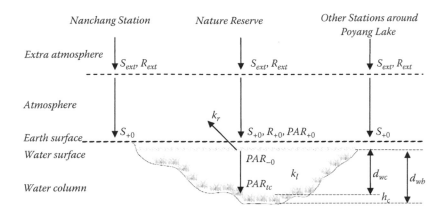

FIGURE 16.4 Model to calculate the light intensity reaching the canopy of SAV. R_{ext}: extra atmosphere global solar radiation; S_{ext}: extra atmosphere or potential sunshine hours; R_{+0}: global solar radiation reaching the earth or water surface; S_{+0}: measured or estimated sunshine hours on the earth or water surface; PAR_{+0}: PAR reaching the earth or water surface; PAR_0: PAR entering water body; PAR_{tc}: PAR reaching the canopy of SAV; k_r: reflection coefficient on water surface; k: light (PAR) attenuation coefficient in water column; d_{wb}: water depth; h_c: height of SAV; d_{wc}: height of water column above canopy.

sunshine hours over the lake. Chen et al. (2006), analyzing the daily global radiation data and sunshine hours from 1994 to 1998 at 86 stations in China, reported that 92% of the daily global radiation could be explained by daily sunshine hours. We applied this model combined with sunshine hours derived through interpolation to estimate daily global solar radiation (R_{+0}) within PLNNR. A possible alternative could be using the duration of cloud cover derived from low-resolution satellite images to predict daily solar radiation (Kandirmaz et al. 2004). We are currently studying the possibility of predicting solar radiation over Poyang Lake from hourly records of cloud cover recorded by the Chinese Fengyun-2C (FY-2C) geostationary meteorological satellite.

The radiation reaching the canopy depends on the light attenuation and depth of the water column (d_{wc}) above the canopy of SAV. The d_{wc} can be derived while using the DEM, measured water levels, and vegetation height. Light attenuation (k) is related to SDD (Figure 16.5), a variable that has been recorded in three lakes within the PLNNR and one lake immediately adjacent to it since 1999. Lillesand and Chipman (2001) mentioned the potential of using low-resolution MODIS satellite data to estimate SDD of large lakes. Lillesand (2004) reported that 79% of variation of weekly to biweekly SDD of eleven inland lakes in Wisconsin could be explained by the ratio of the blue and red MODIS bands. Using the in situ SDD measurements collected in 2004 and 2005 at PLNNR, we developed a regression model between the natural logarithm of SDD and the blue and red bands of MODIS, and it also explained 88% of the SDD variation (Wu et al. 2007a). The SDD predicted from MODIS imagery (Figure 16.6) could thus be used to complement information on water transparency for those lakes where measurements have not been taken. Moreover, it provides a synoptic overview of the variability in transparency within lakes.

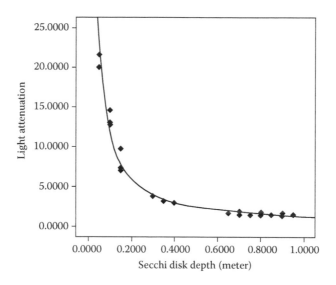

FIGURE 16.5   Relation between light (PAR) attenuation and Secchi disk depth ($R^2 = 0.986$).

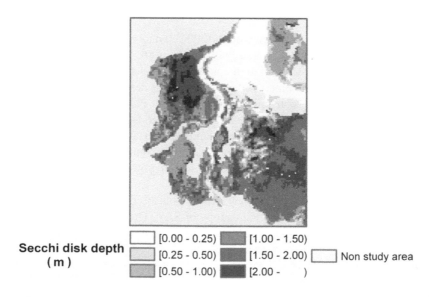

FIGURE 16.6   Secchi disk depths of Poyang Lake National Nature Reserve predicted from a MODIS image dated August 8, 2004. **(See color insert after p. 162.)**

We are implementing the combination of radiation estimates over the lake and SDD with other field-based measurements to calculate the radiation reaching the canopy of vegetation, which could then be used as an input to a SAV simulation model coupled to a GIS.

Another important required variable is water temperature, as it has a pronounced influence on photosynthesis. Photosynthesis over the whole canopy is integrated from

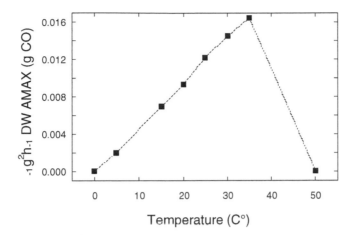

FIGURE 16.7 CO_2 assimilation capacity at light saturation (AMAX), based on data published by E. P. H. Best and W. A. Boyd (2001a). *A simulation model for growth of the submersed aquatic macrophyte American wildcelery (*Vallisneria americana Michx.*).* ERDC/EL TR-01-5. Vicksburg, MS: U.S. Army Engineer Research and Development Center.

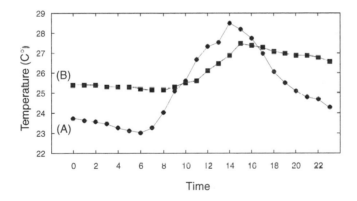

FIGURE 16.8 Average diurnal water temperature from August to September 2005 at different water depths in Poyang Lake.

instantaneous CO_2 gross assimilation rates, and it is a function of absorbed light energy and the photosynthesis light response of individual shoots. As only a small portion of the total incoming solar radiation is used in photosynthesis, growth rates of individual shoots largely depend on the maximum rate of assimilation (AMAX). AMAX is the actual CO_2 assimilation rate at light saturation for individual shoots, which responds dynamically to varying water temperatures at different water depths. For the related species, *Vallisneria americana*, AMAX-to-temperature relations are given in Figure 16.7.

Water temperatures show a diurnal progression, and its amplitude varies with different water depths, as can also be seen from Figure 16.8. Combining AMAX-to-temperature (Figure 16.7) and diurnal temperatures as observed over Poyang Lake (Figure 16.8), it is evident that neither daily average water temperatures from synoptic observations nor observations from polar-orbiting satellites capture diurnal CO_2 gross assimilation rates well. So it is necessary to obtain radiation and temperature data with higher temporal resolution (hourly).

At present, no record of seasonal variation of water temperature within PLNNR exists. RS offers the possibility of measuring surface water temperature. For example, Fox et al. (2005) employed SeaWiFS and AVHRR to estimate the spatial and temporal distributions of water temperature, while Handcock et al. (2006) studied the accuracy and uncertainty of water temperature estimates from the thermal-infrared band of ASTER, MODIS, and Landsat ETM+ images.

Despite the fact that authors have successfully been able to use imagery from polar-orbiting satellites to estimate water temperatures, too few clear-sky observations remain to satisfy our strict model requirements. This gap is due to the relatively low temporal resolution of the images. A challenge lies in the use of geostationary satellite data to estimate water temperature and insolation commensurate with the temporal and radiometric requirements our objective. Fengyun-2C (FY-2C), a geostationary meteorological satellite operated by the General National Satellite Meteorological Center (NSMC) of China, observes a large part of Asia centered at 105 degrees east. Future work is focusing on the development of algorithms for estimating water (skin) temperature and solar irradiance from this relatively new sensor. Theoretically, temperature and radiation can be determined 24 times per day from FY-2C, but in practice, fewer observations are available due to cloud cover. Based on these estimates, an attempt will be made to model underwater temperature and light conditions using an inductive modeling approach as the optimum water depth for the growth of *Vallisneria spiralis* is at 2 to 3 meters depth, and not at the surface (Chambers and Kalff 1987).

The temporal or spatial distribution of other environmental variables, such as total phosphorus and total nitrogen, have also been studied through RS by several researchers (Dewidar and Khedr 2001, Wang and Ma 2001), and their implementation possibilities in Poyang Lake could be explored in the future.

16.6 QUESTIONS TO BE ADDRESSED

A SAV simulation model coupled to a GIS as described above could be used to address important environmental questions, of which three major ones are summarized here.

First, plans are underway to change the hydrology of the Poyang Lake and the Yangtze River. Concern has been expressed that those hydrological changes could negatively affect *Vallisneria* production, and therefore the feeding habitat of wintering Siberian cranes or other tuber-feeding waterfowl. To hydrological engineers and resource planners it would be useful to quantitatively evaluate the impact of future hydrological changes on biota, especially where endangered species, are involved, as in the case at Poyang Lake (Wu and Ji 2002). Development plans could then be redesigned and expensive alterations, once infrastructure has been built, can be avoided.

Second, once engineering projects have been completed, it is important to evaluate the impacts of hydrological changes on the ecosystem in which the submerged aquatic vegetation plays an important role. With the Three Gorges Dam approaching completion, it is not yet understood how various potential water management scenarios from the dam may (or may not) affect SAV production at Poyang Lake, located downstream of the Three Gorges Dam. Predicting impacts from the Three Gorges Dam is difficult because there is a tremendous seasonal and year-to-year hydrological variability. Dredging is another activity changing the hydrology of Poyang Lake (Wu et al., 2007b). It has been suggested that it might influence the prouduction of *Vallisneria,* but the exact impacts remain to be assessed. SAV-dominated ecosystems of Poyang Lake likewise respond to hydrological variability (International Crane Foundation and PLNNR, unpublished data). SAV simulation models, like VALLA, might augment our understanding of the system behavior and allow us to assess the most important variables to monitor. A calibrated model predicting tuber biomass production, as affected by hydrology, will also enable us to distinguish between future changes in tuber biomass attributed to natural seasonal and year-to-year variations versus man-made hydrological changes.

A third application of a SAV simulation model could be to evaluate whether tuber-feeding bird populations are constrained by the available food sources and in what years. Such an analysis can only be undertaken at the level of the catchments within which the birds reside. Application of GIS-based SAV models might ultimately enable conservationists to address such questions.

ACKNOWLEDGMENTS

Financial support for this research was provided by the Stichting voor Wetenschappelijk Onderzoek van de Tropen (the Netherlands Foundation for the Advancement of Tropical Research [WOTRO], File number: WB 84-550) and the International Institute for Geo-information Science and Earth Observation (ITC), the Netherlands. We also acknowledge support from the School of Resource and Environmental Sciences, Wuhan University, China; the International Crane Foundation, United States; the Jiangxi Bureau of Forestry, China; and the Bureau of Jiangxi Poyang Lake National Nature Reserve, China.

REFERENCES

Bartell, S. M., et al. 2000. *Interim report for the upper Mississippi River system—Illinois waterway system navigation study. Ecological risk assessment of the effects of incremental increase of commercial navigation traffic (25, 50, 75, and 100% increase of 1992 baseline traffic) on submerged aquatic plants in the main channel borders.* ENV Report 17. Rock Island, IL: U. S. Army Corps of Engineers, Rock Island District, St. Louis District, St. Paul District.

Best, E. P. H., and W. A. Boyd. 2001a. *A simulation model for growth of the submersed aquatic macrophyte American wildcelery (Vallisneria americana Michx.).* ERDC/EL TR-01-5. Vicksburg, MS: U.S. Army Engineer Research and Development Center.

Best, E. P. H., and W. A. Boyd. 2001b. *VALLA (version 1.0): A simulation model for growth of American wildcelery.* ERDC/EL SR-01-1. Vicksburg, MS: U.S. Army Engineer Research and Development Center.

Best, E. P. H., et al. 2004. *Modeling the impacts of suspended sediment concentration and current velocity on submersed vegetation in an Illinois river pool, USA.* ERDC/TN APCRP-EA-07. Vicksburg, MS: U.S. Army Engineer Research and Development Center.

Best, E. P. H., et al. 2005. *Aquatic plant growth model refinement for the upper Mississippi River—Illinois waterway system navigation study.* ENV Report 51. Rock Island, IL: U. S. Army Corps of Engineers, Rock Island District, St. Louis District, St. Paul District.

Calado, G., and P. Duarte. 2000. Modelling growth of *Ruppia cirrhosa. Aquatic Botany* 68:29–44.

Carr, G. M., et al. 1997. Models of aquatic plant productivity: A review of the factors that influence growth. *Aquatic Botany* 59:195–215.

Chambers, P. A., and J. Kalff. 1987. Light and nutrients in the control of aquatic plant community structure. *Journal of Ecology* 75:611–619.

Chen, R., et al. 2006. Estimating daily global radiation using two types of revised models in China. *Energy Conversion and Management* 47:865–878.

Collins, C. D., and J. H. Wlosinski. 1989. A macrophyte submodel for aquatic ecosystems. *Aquatic Botany* 33:191–206.

Dewidar, K., and A. Khedr. 2001. Water quality assessment with simultaneous Landsat-5 TM at Manzala Lagoon, Egypt. *Hydrobiologia* 457:49–58.

Fox, M. F., et al. 2005. Spatial and temporal distributions of surface temperature and chlorophyll in the Gulf of Maine during 1998 using SeaWiFS and AVHRR imagery. *Marine Chemistry* 97:104–123.

Giusti, E., and S. Marsili-Libelli. 2005. Modelling the interactions between nutrients and the submersed vegetation in the Orbetello Lagoon. *Ecological Modelling* 184:141–161.

Handcock, R. N., et al. 2006. Accuracy and uncertainty of thermal-infrared remote sensing of stream temperatures at multiple spatial scales. *Remote Sensing of Environment* 100:427–440.

Harwell, M. C., and K. E. Havens. 2003. Experimental studies on the recovery potential of submerged aquatic vegetation after flooding and desiccation in a large subtropical lake. *Aquatic Botany* 77:135–151.

Herb, W. R., and H. G. Stefan. 2003. Integral growth of submersed macrophytes in varying light regimes. *Ecological Modelling* 168:77–100.

Hootsmans, M. J. M. 1994. A growth analysis model for *potamogeton pectinatus* L. In *Lake Veluwe, a macrophyte-dominated system under eutrophication stress*, ed. W. Van Vierssen, M. Hootsmans, and J. Vermaat. Dordrecht, the Netherlands: Kluwer Academic Publishers, 250–286.

Huang, B., and B. Jiang. 2002. AVTOP: A full integration of TOPMODEL into GIS. *Environmental Modelling and Software* 17:261–268.

Jahn, L. A., and R. V. Anderson. 1986. The ecology of pools 19 and 20, of the upper Mississippi River: A community profile. U.S. Fish and Wildlife Service, Biological Report 8517.67.

Jonzen, N., et al. 2002. Seasonal herbivory and mortality compensation in a swan-pondweed system. *Ecological Modelling* 147:209–219.

Kanai, Y., et al. 2002. Migration routes and important resting areas of Siberian cranes (*Grus Leucogeanus*) between northeastern Siberia and China as revealed by satellite tracking. *Biological Conservation* 106:339–346.

Kandirmaz, H. M., et al. 2004. Daily global solar radiation mapping of Turkey using metepsat satellite data. *International Journal of Remote Sensing* 25:2159–2168.

Korschgen, C. E. 1989. Riverine and deepwater habitats for diving ducks. In *Habitat management for migrating and wintering waterfowl in North America*, ed. L. M. Smith, R. L. Pederson, and R. M. Kaminski. Lubbock: Texas Tech University Press, 157–180.

LaMontagne, J. M., et al. 2003. Compensatory growth responses of *Potamogeton pectinatus* to foraging by migrating trumpeter swans in spring stop over areas. *Aquatic Botany* 76:235–244.

Lillesand, T. M. 2004. Combining satellite remote sensing and volunteer Sechhi disk measurement for lake transparency monitoring. http://www.nwqmc.org/NWQMC-Proceedings/Papers-Alphabetical%20by%20First%20Name/Thomas%20Lillesand-Satellite.pdf (accessed June 8, 2006).

Lillesand, T. M., and J. W. Chipman. 2001. Satellite-assisted lake water quality: Using satellite data to observe regional trends in lake transparency. *GIM International* 15:26–29.

Meine, C. D., and G. W. Archibald, eds. 1996. The cranes—status survey and conservation action plan. Gland, Switzerland: IUCN.

Scheffer, M., et al. 1993. MEGAPLANT: A simulation model of the dynamics of submerged plants. *Aquatic Botany* 45:341–356.

Van Nes, E. H., et al. 2003. Charisma: A spatial explicit simulation model of submerged macrophytes. *Ecological Modelling* 159:103–116.

Van Vierssen, W., et al. 1994. *Lake Veluwe, a macrophyte-dominated system under eutrophication stress.* Dordrecht, the Netherlands: Kluwer Academic Publishers.

Wang, X. J., and T. Ma. 2001. Application of remote sensing techniques in monitoring and assessing the water quality of Taihu Lake. *Bulletin of Environmental Contamination and Toxicology* 67:863–870.

Wu, G., J. De Leeuw, A. K. Skidmore, H. H. T. Prins, and Y. Liu, (2007a). Comparison of MODIS and Landsat TM5 images for mapping tempo-spatial dynamics of Secchi disk depths in Poyang Lake national nature reserve, China, *International Journal of Remote Sensing,* In Press, Accepted Manuscript.

Wu, G., J. De Leeuw, A. K. Skidmore, H. H. T. Prins, and Y. Liu, (2007b). Concurrent monitoring of vessels and water turbidity enhances the strength of evidence in remotely sensed dredging impact assessment. *Water Research,* 41(15):3271–3280.

Wu, Y., and W. Ji. 2002. *Study on Jiangxi Poyang Lake National Nature Reserve.* Beijing, China: Forest Publishing House.

17 Soundscape Characteristics of an Environment
A New Ecological Indicator of Ecosystem Health

*Jiaguo Qi, Stuart H. Gage, Wooyeong Joo,
Brian Napoletano, and S. Biswas*

17.1 INTRODUCTION

Landscape characteristics are important measures of an ecosystem's environmental health, as they depict spatial patterns of physical attributes of the landscape that many organisms rely on. The visual features of a landscape, such as forest type, density, patch size and shape, affect habitat properties that are specific to different organisms. Change or disruption of the spatial patterns of a landscape has been shown to impact biodiversity (Crist et al., 2004, Jeanneret et al., 2003, Sala et al., 2000, Foley et al., 2005), ecological function (Allan, 2004, Alberti, 2005, Grigulis et al., 2005, Battin, 2004), and ecosystem services (Tscharntke et al., 2005, Fischer and Lindenmayer, 2007).

A suite of landscape matrices has been developed based on land use and land cover maps derived from satellite images as a measure of landscape fragmentation. They include, for example, patch density, Shannon diversity index, as proxies of landscape characteristics. These matrices have been found to be important indicators of an ecosystem's biodiversity and integrity (Sala et al., 2000, Foley et al., 2005, Fischer and Lindenmayer, 2007).

Although these landscape characteristics, often derived from analysis of remotely sensed imagery, are important indicators of ecosystem health, they are temporally static and do not provide a sufficient spatial resolution to observe the responses of individual organisms to anthropogenic disturbances. The audio characteristics emitted from an ecosystem, such as sounds from birds, mechanical movements, or wind (Truax, 1999, Schafer, 1977), provide unique insight into spatial and temporal patterns of ecosystem responses to human disturbances. While soundscape characteristics provide complementary information to landscape characteristics, little research has been done to fully explore the usefulness of coupling these two complementary indicators of ecological dynamics.

We define an ecosystem's soundscape as the physical extent of acoustic signals and the spectral range of signal frequencies associated with an ecosystem's biophysical processes. Truax (1999) and Schafer (1977) introduced the idea of a soundscape in their early studies of acoustic ecology. Environmental soundscape analysis as a complementary measure of ecosystem dynamics uniquely addresses some of the key criteria for the establishment of ecological indices as articulated by Dale and Beyler (2001). Soundscape analysis is a predictable measure of ecosystem stress, is anticipatory, is integrative, and can measure disturbance. Because an ecosystem's soundscape is a function of a variety of ecological variables, assessment of the soundscape serves to integrate several variables in the measure of integrity and biocomplexity (Thompson 2001, Holling 2001, Mueller and Kuc 2000, Porter et al., 2005). This chapter demonstrates the capability of acoustic sensing techniques to characterize an ecosystem's soundscape.

17.2 ACOUSTIC SIGNAL CLASSIFICATION

Viewed in terms of information theory, the acoustic frequency spectrum is primarily an information-carrying medium. An organism or force generating the acoustic signal acts as the encoder and transmitter, and the acoustic spectrum acts as the medium through which the encoded signal travels. The receiver then registers and decodes the signal (as in human conversation, for instance). The various signals within the acoustic spectrum are commonly classified as either natural or human-induced sounds (Schafer 1977).

Krause (1998), in his studies of natural soundscapes, devised the term *biophony* to describe the complex chorus of ambient biological sounds (biophony = biologic and symphony), and *geophony* for a region's ambient geological sounds (Figure 17.1). Similarly, the term *anthrophony* refers to the human-imposed sounds (0.2–2.0 kHz). The two primary categories, biophony and anthrophony, can be further subdivided conceptually. Early observations led to the conclusion that signals within the biophony range (2.0–11.0 kHz) can be characterized as intentional, meaning the transmitter of the signal wishes to communicate information, such as mating or distress calls, through the acoustic spectrum, or incidental, in which signals transmitted may contain relevant information but are not dispatched for the explicit purpose of communication.

Anthropogenic sounds can be further divided into mechanistic and oral classes. Oral sounds are those produced by human beings themselves (i.e., talking, shouting, or singing). Conversely, mechanistic signals involve sounds produced by various forms of human-made machinery and technology. Within this class, two further subcategories exist: stationary and temporal. Stationary refers to those signals that impose themselves on the ambient soundscape permanently (i.e., turbulence from air-conditioner fans), and temporal signals include the noises that move through the soundscape over a given temporal scale (i.e., automobile or train traffic). While this schema does not provide an absolute standard of acoustic classification, it does provide the framework to begin characterization of acoustic signaling (see Figure 17.1).

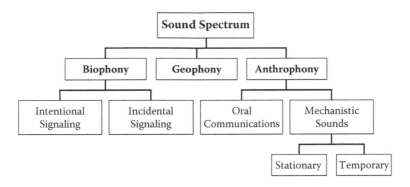

FIGURE 17.1   View diagram of acoustic taxonomy.

17.3 SOUNDSCAPE ANALYSIS

17.3.1 ECOLOGICAL SOUNDSCAPES

Acoustic diversity refers to the patterns of frequency and temporal use of the acoustic spectrum. Biophonic complexity thereby indicates the degree to which different vocalizing organisms utilize different niches to relay information within the spectrum. Specifically, ecosystems with lesser degrees of human interference tend to exhibit greater biophonic complexity in terms of frequency and periodicity utilization. Moreover, anthropogenic interference, and more particularly temporal interference, within a soundscape will tend to hinder organism populations by lowering reproduction rates and increasing predation rates. Organisms make careful use of the acoustic frequency when attempting to communicate information such as mating potential, territory size, and potential predation. When anthropogenic interference disrupts this communication, critical information is not relayed and the organism's population experiences a decline (Krause 1998). Therefore, acoustic characteristics may serve as an ecological indicator of ecosystems.

17.3.2 DEVELOPMENT OF SOUNDSCAPE INDICATORS

An acoustic signal is characterized by multiple physical attributes including timing, frequency, and intensity. The data set produced by acoustic recordings and quantification is an array of acoustic intensity of contiguous, nonoverlapping frequency bands (Figure 17.2). These data form a data matrix where the rows represent recording intervals and the columns are frequency bands. A wide frequency band summarizes the intensity of sound waves across a relatively wide set of frequencies, while a narrow band restricts the range of frequency summarized. The analytical role is to summarize patterns in covariation among the different frequency bands across the temporal period during which the acoustic data were recorded. The most convincing and feasible statistical method for describing such patterns of covariation in each acoustic signature is to calculate the dominance in each frequency band and compute their statistical distributions.

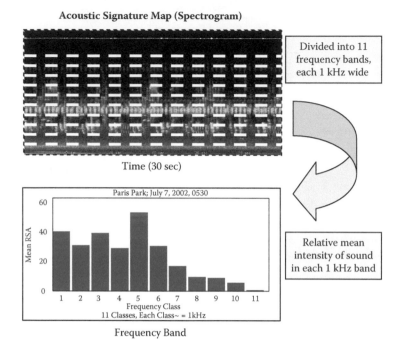

FIGURE 17.2 The acoustic frequency slicing procedure. Each sound wave file is divided into 11 frequency bands and the relative mean intensity is calculated for each band. **(See color insert after p. 162.)** Note that the 5-kHz band has the highest mean intensity across the 11 frequency bands.

In impacted ecosystems the spectral properties of acoustic signals in the environment sometimes aggregate within two primary regions of a spectrogram. The first region occurs at the lower frequencies of the sound spectrum. This band typically extends from 0.2 to 1.5 kHz and consists primarily of mechanical signals (e.g., trains, cars, air conditioners, etc.), and is therefore referred to as the *anthrophonic region*. The second band of concentration begins in the range of 2 kHz and is prevalent up to 8 kHz, but may reach a higher spectral range especially when organisms communicate using wider signal bandwidths (e.g., *Molothrus ater*) or ultrasound (e.g., bats). We currently restrict our range to human detection to match with human auditory survey techniques. This realm of acoustic activity consists primarily of signals generated by biological organisms, and is therefore referred to as the *biophonic region*. We have delineated this frequency band as the biological band based on our observations and the frequency ranges referred to in the literature. These two bands correspond to two of the three taxonomic categories of the soundscape described above, but do not cover acoustics emanating from the physical (i.e., wind, rain, etc.) or geophonic component. This is because the geophony, when present, occurs as a signal that is diffuse throughout the entire spectrum. The geophony is a diffuse signal that is strongest at the lowest frequencies, but continues with a relatively high intensity into the higher frequencies, and its individual components are difficult to isolate and identify.

Using this structure we compute the acoustic intensity for anthrophony (α), biophony (β), and geophony (γ). These three acoustic ranges are then compared to the value of the entire signal (σ). A value > 1 indicates that the concentration of acoustic activity in the analyzed region was greater than the value for the entire signal. Therefore, the region with the highest value was the predominant source of acoustic activity in the signal. For example, if the β_r had the highest value, then biological activity was predominant, while a larger α_r value indicated dominant anthropogenic activity. To emphasize the comparison of biological and anthropogenic activity, we divided the β value by the α value to calculate ρ (=β/α), the ratio of biological to anthropogenic activity.

In addition to computing the ratios of activity from our spectrumgram, we also determined the percentage of total activity a single band contributes to the total signal. A γ_p value near 100% coincident with a β_p value of approximately the same value indicated that the primary signal source in the sound sample was *biophony* (geophysical) activity. When the α_p value was greater than 50%, it indicated that the primary signal source was *anthrophony* (anthropogenic) activity, whereas a value of β_p greater than 50% indicated that *biophony* (biological) activity was the dominant source.

17.4 A SAMPLE APPLICATION

To demonstrate the usefulness of the acoustic signals as an environmental indicator, sounds were recorded in Nanchang city, China (Figure 17.3) and another one in Michigan. Nanchang Park was once a plant nursery but was transformed into a

FIGURE 17.3 A photograph of the China study site where acoustic data were collected and analyzed in this paper.

natural reserve after it changed owners in 1996. Soon after that, the park became one of the primary nesting and mating areas for summer migratory birds. Sound recording ecosystems were developed and calibrated, and the sounds were recorded between July 7 and 15, 2005 at 30-minute time intervals. The Michigan site was located in a backyard of a private house in a rural residential area in Okemos, Michigan, surrounded by forests woodlots. Acoustic recorders were placed about 40 yards away from the house for a multiple year data collection. However, in this study, we only used a short period of time data in July 7, 2005 that are coincident with the data from China.

As a demonstration of the soundscape characteristics, Figure 17.4, depicts the sound spectra of selected acoustic signals from data collected on July 5, 2005 at 7:30 p.m. local time in Nanchang (top) and on July 9, 2005 at 5:30 a.m. in Michigan (bottom). The horizontal axis is the time (30 seconds in this case) while the y-axis is the frequency. The brightness of the image represents the vocal strength or intensity. The brighter the image, the intense or loud the sound is. The two spectra from Michigan and Nanchang showed different acoustic patterns suggesting different biological activities at the two sites.

The two sites also showed different proportions of biological and anthropogenic activities. Analysis of the acoustic signals in the frequency domain (Figure 17.5) suggest that Michigan site had more biological signals than anthropogenic activities while the Nanchang site has almost equal biological and anthropogenic activities, as indicated in the histograms of the frequency. Although qualitative, the Nanchang site indeed had more human related acoustic signals as it is in the Center of the big city, Nanchang, China, while the site in Michigan is a residential area at the outskirts of a small city, Okemos, Michigan. The ratios of biological to anthropogenic signals ($\rho = \beta/\alpha$) of the two sites are compared in Figure 17.6 and they suggest the same results as in Figure 17.5 that the biological activities are dominant at the Michigan site while the anthropogenic activities were dominant at the Nanchang site.

Another type of application of the acoustic sensing technology is monitoring of bird species through pattern recognition. Once an acoustic image is generated, a signature of a specific bird, for example, can be identified (Figure 17.7). This identified acoustic signature (training signature) can then be used in image processing to search for similar patterns in other acoustic data, thus detecting the presence of such bird. Once expanded in time series, one can detect and monitor bird species and possibly population.

17.5 DISCUSSION AND CONCLUSIONS

The research results presented in this paper represent a frontier work in expanding traditional remote sensing to acoustic sensing. The fundamental difference between traditional remote sensing and acoustic remote sensing is that the former utilizes electromagnetic fields while the latter relies on air for signal transmission. Therefore, a series of questions arises that needs to be addressed. The first one is related to the transmission of acoustic signals—how far does the acoustic signal travel, that is, what is the distance between the recording device and the sound of origin? This may well depend on the location of the sensor (in forested lands, grasslands, open

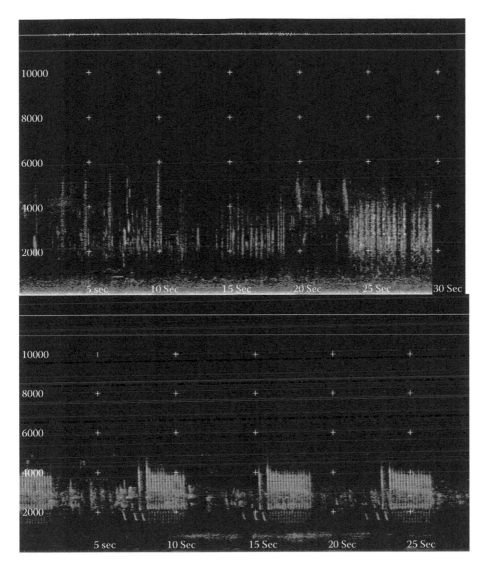

FIGURE 17.4 Sound spectra of selected acoustic signals from data collected on July 5, 2005 at 7:30 p.m. local time in Nanchang (top) and on July 9, 2005 at 5:30 a.m. in Michigan (bottom). (**See color insert after p. 162.**)

urban lands) and its surrounding physical environment. One may record the acoustic signal of a bird, for example, but may also realize that the bird was just flying over rather than inhabiting the landscape where the sensor is placed. Unlike traditional remote sensing where each pixel is associated with a fixed physical dimension of a landscape (e.g., pixel size), acoustic signals do not have a fixed range of physical dimension, as the recorded signals will vary depending on the sensor's sensitivity, distance of sound of origin, and physical characteristics of the environment (windy days, or densely forested environment, for example). Therefore, interpretation of

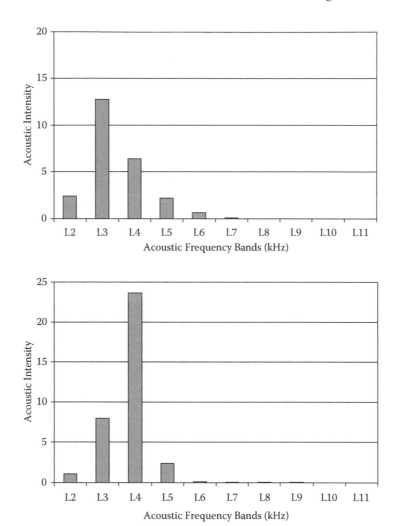

FIGURE 17.5   Frequency distributions of the acoustic spectra from Figure 17.4.

acoustic signals is best achieved when considering the physical environment or land-scape properties.

The use of acoustic signals as an ecological indicator is only feasible for infer-ring ecological information of those species that generate vocal signals. Amphib-ians and mammals, for example, do not generate sounds that can be recorded with traditional recording devices. Thus, at this time, we can only infer information about vocal species.

The temporal characteristics of acoustic signals are critical components of any interpretation. Unlike the physical environment of a landscape, the soundscape is a very dynamic field that varies considerably within a short period of time. Diurnal behavior of many bird species would result in a strong biological frequency in a soundscape in the early morning, while crickets are active in the evening. These

The Indices of Environmental Acoustic Signals

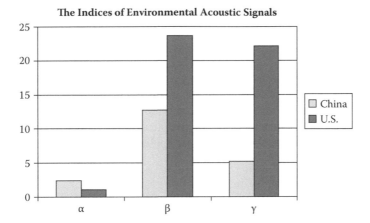

FIGURE 17.6 Calculated alpha (α), beta (β), and their ratios using the data from Figure 17.5.

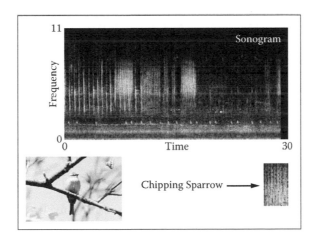

FIGURE 17.7 Demonstration using acoustic signals in time series analysis to identify bird species and population. (See color insert after p. 162.)

temporal characteristics need to be considered when attempting to capture the biological soundscape of these species.

The analytical methods used in this paper are only examples in analyzing acoustic signals and there are other ecological indicators that can be derived from acoustic signals. However, this paper represents the first involving remote sensing that utilizes frequencies or wavelengths that can only be transmitted through a physical medium such as air. Nevertheless, the expansion of the remote sensing concept to acoustic signal analysis has provides complementary and useful information about the ecological characteristics of an environment. When applied spatially and temporally across a landscape, much more comprehensive information can be inferred. For example, a network of sensors in a city with simultaneous measurements of acoustic signals may provide not only information on ecological characteristics, but also a quantitative measure of human-induced noise levels across the entire city, which is a very valuable indicator of the environmental quality of the city. With long-term measurements of such acoustic signals, one may further understand environmental degradation processes.

Finally, this technology is relatively inexpensive compared with traditional remote sensing devices, and therefore can be deployed to obtain long-term and spatially distributed data. Furthermore, the operation of recording devices is relatively simple and inexpensive in comparison with optical remote sensing devices, thus providing a convenient technology for broader applications.

ACKNOWLEDGMENTS

The Great Lakes Fisheries Trust provided support for investigating acoustic signals as part of a grant entitled *Ecological Assessment of the Muskegon River Watershed* awarded to a consortium of investigators. This work was also supported by the NASA grant (NNG05GD49G) and by a grant at IGSNRR of Chinese Academy of Sciences (Human Activities and Ecosystem Changes). We want to thank Nathan Torbick for installation of the recording devices and data recording, Liu Ying at Jiangxi Normal University for his assistance in data acquisition, and Weitao Ji at the Poyang Lake Station for allowing the authors to use their facilities at Tiangxing Yuan Park and Poyang Lake.

REFERENCES

Alberti, M., 2005, The effects of urban patterns on ecosystem function. *International Regional Science Review* Vol. 28, No. 2, 168–192

Allan, J. David, 2004, Landscapes and riverscapes: the influence of land use on stream ecosystems. *Annual Review of Ecology, Evolution, and Systematics* Vol. 35:257–284

Battin, J., 2004, When good animals love bad habitats: Ecological traps and the conservation of animal populations. *Conservation Biology* 18, 1482–1491.

Crist, P. J. , T. W. Kohley and J. Oakleaf, 2004. Assessing land-use impacts on biodiversity using an expert systems tool. *Landscape Ecology* Vol. 15, no. 1, pp. 1–84.

Dale, V. H., and S. C. Beyler. 2001. Challenges in the development and use of ecological indicators. *Ecological Indicators* 1:3–10.

Fischer, Joern and David B. Lindenmayer, 2007. Landscape modification and habitat fragmentation: A synthesis. *Global Ecology and Biogeography* 16 (3), 265–280.

Foley, J.A., R. DeFries, G.P. Asner, C. Barford, G. Bonan, S.R. Carpenter, F.S. Chapin, M.T. Coe, G.C. Daily, H.K. Gibbs, J.H. Helkowski, T. Holloway, E.A. Howard, C.J. Kucharik, C. Monfreda, J.A. Patz, I.C. Prentice, N. Ramankutty, and P.K. Snyder, 2005. Global consequences of land use. *Science* 309, 570–574.

Grigulis, Karl, Sandra Lavorel, Ian D. Davies, Anabelle Dossantos, Francisco Lloret, Montserrat Vilà, 2005. Landscape-scale positive feedbacks between fire and expansion of the large tussock grass, *Ampelodesmos mauritanica* in Catalan shrublands. *Global Change Biology* 11(7), 1042–1053.

Holling, C. S. 2001. Understanding the complexity of economic, ecological, and social systems. *Ecosystems* 4:390–405.

Jeanneret P., B. Schüpbach, H. Luka, and W. Büchs 2003. Quantifying the impact of landscape and habitat features on biodiversity in cultivated landscapes. *Biotic indicators for biodiversity and sustainable agriculture* 2003, Vol. 98, no. 1–3, pp. 311–320.

Kime, N. M., W. R. Turner, and M. J. Ryan. 2000. The transmission of advertisements calls in Central American frogs. *Behavioral Ecology* 11:71–83.

Krause, B. 1998. *Into a wild sanctuary: A life in music and natural sound.* Berkeley, CA: Heyday Books.

Mueller, R., and R. Kuc. 2000. Foliage echoes: A probe into the ecological acoustics of bat echolocation. *Journal of the Acoustical Society of America* 108:836–845.

Naguib, M. 1996. Ranging by song in Carolina Wrens Thryothorus ludovicianus: Effects of environmental acoustics and strength of song degradation. *Behaviour* 133:541–559.

Penna, M. a. R. S. 1998. Frog call intensities and sound propagation in the South American temperate forest region. *Behavioral Ecology and Sociobiology* 42:371–381.

Porter, J., P. Arzberger, H.W. Braun, P. Bryant, S. Gage, T. Hansen, P. Hanson, C.C. Lin, F. P. Lin, T. Kratz, W. Michener, S. Shapiro, and T. Williams, 2005. *BioScience* Vol. 55, no. 7, pp. 561–572

Sala, O. E., F.S. Chapin, J.J. Armesto, E. Berlow, J. Bloomfield, R. Dirzo, E. Huber-Sanwald, L.F. Huenneke, R.B. Jackson, A. Kinzig, R. Leemans, D.M. Lodge, H.A. Mooney, M. Oesterheld, N.L. Poff, M.T. Sykes, B.H. Walker, M. Walker, and D.H. Wall. 2000. Global biodiversity scenarios for the year 2100. *Science* 287, 1770–1774.

Schafer, R. M. 1977. *The soundscape: Our sonic environment and the tuning of the world.* Rochester, VT: Destiny Books.

Skole, D., and C. Tucker. 1993. Tropical deforestation and habitat fragmentation in the Amazon: Satellite data from 1978 to 1988. *Science* 260:1905–1910.

Snedden, W. A., M. D. Greenfield, and Y. Jang. 1998. Mechanisms of selective attention in grasshopper choruses: Who listens to whom? *Behavioral Ecology and Sociobiology* 43:59–66.

Thompson, J. N. 2001. Frontiers of ecology. *BioScience* 51:15–24.

Truax, B. 1999. *Handbook for acoustic ecology.* Burnaby, BC: Cambridge Street Publishers.

Tscharntke, T., A. M. Klein, A, Kruess, I. Steffan-Dewenter, and C. Thies, 2005. Landscape perspectives on agricultural intensification and biodiversity - ecosystem service management. *Ecology Letters* 8 (8), 857–874.

Part V

Watershed Assessment
and Management

18 Geospatial Decision Models for Assessing the Vulnerability of Wetlands to Potential Human Impacts

Wei "Wayne" Ji and Jia Ma

18.1 INTRODUCTION

Characterized by a shallow water table (Sharitz and Batzer 1999), wetlands are transitional landscapes between open water systems and terrestrial uplands. They provide many crucial ecosystem functions and values, such as flood control, groundwater recharge, sediment and pollutant retention/stabilization, nutrient removal/transformation, and fish and wildlife habitat and diversity (Mitsch and Gosselink 2000). Wetlands are prone to be filled in, drained, or ponded for a variety of human uses including stream channelization and maintenance, urban development, transportation improvement, or conversion to agricultural uses (Dodds 2002). To protect wetland resources, many laws and regulatory programs have been established. Among them, the Clean Water Act (CWA) Section 404 is the primary federal law aiming to maintain and restore the chemical, physical, and biological integrity of the wetlands in the United States. It authorizes U.S. federal agencies, mainly the U.S. Army Corps of Engineers, to issue permits for the discharge of dredged or fill material into the navigable waters at specific disposal sites, including wetlands (USACE: http://www.usace.army.mil/cw/cecwo/reg/sec404.htm). To comprehensively evaluate individual or cumulative impacts of human activities on existing wetlands, a regulatory permit assessment requires quickly retrievable environmental and socioeconomic data, and more importantly, a scientifically justifiable evaluation framework for analyzing those data. In recent decades, GIS techniques have been increasingly used to facilitate the data management and visualization in regulatory wetland assessments or permit reviews, aiming to improve the efficiency of the permit assessment process. A pilot decision supporting GIS for the permit analysis was developed in the early 1990s (Ji and Johnston 1994, 1995). The system was based on a widely used commercial GIS (Arc/Info, ESRI, Inc.), with customized user interfaces for data retrieval, visualization, and analysis. Other similar GIS-based technical tools were also developed, such as the Permit Application Management System (PAMS) for evaluating and tracking the status of permit applications submitted for approval by the Connecticut Department of Environmental Protection, and ERATools for managing permit data and analyzing potential impacts of permitted activities by the Florida Department of Environmental Protection. However, while these GIS tools are powerful and useful in data management, visualization, and spatial analysis, they usually lack the decision rules or models that can link geospatial data manipulations to evaluating how vulnerable wetland functions and values would be under potential human impacts in the context of regulatory assessment. Thus, rule-based decision models need to be developed and incorporated with the GIS tools.

During recent decades, numerous environment assessment models have been developed, which can be used, at least partially, to evaluate wetland functions and values for various decision-making purposes. Examples of these models include the Habitat Evaluation Procedure (HEP) (USFWS [U.S. Fish and Wildlife Service] 1980), the Wetland Evaluation Technique (WET) (Adamus 1983), the Index of Biological Integrity (IBI) (Karr 1997), the GIS-based Wetland Value Assessment Methodology (Ji and Mitchell 1995), and the Hydrogeomorphic Approach (HGM) (Hollands and Magee 1985, Brinson 1996). However, none of the existing models can be effectively used with GIS data to assess wetlands for regulatory wetland assessments, such as the Section 404 permit review. This is because (1) these models were originally developed for other applications (e.g., wetland restoration planning or wildlife habitat evaluation), and thus do not address all the functions, such as socioeconomic function, of wetlands that need to be assessed in regulatory assessments; and (2) all of these models require a great amount of field data collection and specialized expertise for implementation. Therefore, it is not effective and efficient to directly adopt and integrate the existing models with GIS to address the needs in wetland regulatory assessment. Clearly, there is a critical need for GIS-based decision models in order to handle increasing volumes of existing geospatial data for rapidly assessing wetland vulnerability in management decision making (USEPA 2004 research solicitation: EPA FRL-OW-7620-6).

To address this objective, our efforts focused on the design of geospatial decision models that generate a ranked wetland vulnerability index (WVI) based on geospatial data and analysis. In addition, a user-friendly decision support GIS with customized user interfaces was developed to facilitate the implementation of the models. The developed decision models were applied to the Little Blue River watershed in the state of Missouri in the United States.

18.2 GEOSPATIAL DECISION MODELS

A geospatial decision model is one that generates output for management decision support, such as ranked indices, based on geospatial data and analysis. In this study, the models are to be used for assessing wetland vulnerability, which is defined as the degradation likelihood of wetland functions and values under potential anthropogenic pressures. Certain characteristics of wetlands (e.g., the size or recreational usage) and the spatial occurrence of certain concerned entities (e.g., endangered species or a historical site) related to particular wetlands may largely determine the degree of vulnerability of the wetland's functions and values. Thus, the geospatial

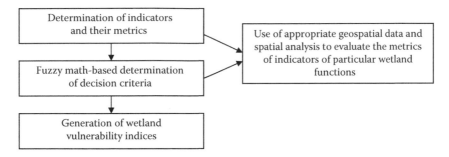

FIGURE 18.1   Wetland vulnerability assessment with geospatial decision models.

decision models were developed so they can be used to identify and evaluate these characteristics and the concerned entities of wetlands with GIS-based data and analysis; they follow a 3-step procedure (Figure 18.1):

18.2.1 DETERMINATION OF INDICATORS AND METRICS

To address the fundamental needs of the regulatory assessment of wetlands, four wetland functions were selected for our modeling: (1) biological supporting function, (2) hydrological function, (3) physiographic function, and (4) socioeconomic function. For each of these functions, three indicators were identified (Table 18.1). The selection of the indicators for the geospatial decision models follows three considerations:

(a) the selected indicators of a particular wetland function should be able to address major concerns of wetland regulatory assessment;
(b) the selected indicators should have been used in wetland assessment by related environmental management agencies or identified in research publications; and
(c) the indicators can be evaluated using GIS-based data and analysis.

The indicator selection process involved consultations with related governmental agencies that are responsible for wetland regulatory assessment, including the U.S. Army Corps of Engineers (USACOE) and the U.S. Environmental Protection Agency. We also conducted literature reviews (e.g., USEPA 2002, Stein 1998, Adamus 1983, USACOE 1997, Hollands and Magee 1985, Brinson 1996, Sousa 1985, Cook et al. 1993, Karr and Chu 1998, Zampella 1994, Hruby et al. 1995). As shown in Table 18.1, the measurement of an indicator is referred to as *metrics*. According to the value of the metrics, the decision criteria (or decision boundaries) for evaluating a particular indicator are determined. Ranking scores are assigned based on the decision criteria. When necessary, weights may be determined and applied to certain ranking scores. To evaluate a particular indicator, appropriate GIS data and spatial analysis operations need to be identified and used, as shown in the last column of Table 18.1. The metrics are either geospatial or descriptive. Geospatial metrics can measure the spatial occurrence and size of a wetland or concerned entities in a potentially impacted area, or their spatial proximity to concerned human activity locations. The descriptive metrics help identify certain characteristics or features of

TABLE 18.1

Structure of geospacial decision models. The indicators and metrics are for selected wetland functions. The decision criteria and score/weight columns show the examples of possible metric values and scores. The GIS data column illustrates some typical data sets that can be used to evaluate the metrics.

Indicator	Metrics	Decision criteria	Score	Weight	GIS data
Biological supporting function					
Total area of a target wetland (BV1)	Total area of a wetland in a potentially impacted location	>75 percentile ≥25 and <75 percentile <25 percentile	1.0 0.5 0.1		Wetland data and maps
Proximity to species of concern (BV2)	The number of species of concern	>5 species ≥2 and <5 species <2 species	1.0 0.5 0.1	3 2 1	Wildlife species data
Habitat fragmentation (BV3)	The density of road	>20% ≥10% and <20% <10%	1.0 0.5 0.1		Road density data
Hydrological function					
Flood risk (HV1)	The percentage of floodway	>75 percentile ≥25 and <75 percentile <25 percentile	1.0 0.5 0.1		Stream/river data for high-risk flooding regions
Hydrological modification (HV2)	The occurrence of dams	Occurrence Not occurrence	1 0		Datasets for dams

Criterion	Description	Category	Value	Data source
Pollution potential (HV2)	The number of pollution sites	>75 percentile ≥25 and <75 percentile <25 percentile	1.0 0.5 0.1	Landfills and mining wastes data, etc.
Physiographic function				
Erosion potential (PV1)	The percentage of erodible soil	>75 percentile ≥25 and <75 percentile <25 percentile	1.0 0.5 0.1	Soil data and maps
Drinking water relevance (PV2)	The occurrence of public water supply facilities	Occurrence Not occurrence	1 0	PWS lakes, tanks, wells, and springs data
Nearby land uses (PV3)	The percentage of urban and agricultural lands	>75 percentile ≥25 and <75 percentile <25 percentile	1.0 0.5 0.1	Land use/land cover data
Socioeconomic function				
Proximity to important public land (SV1)	Percentage of public lands	>75 percentile ≥25 and <75 percentile <25 percentile	1.0 0.5 0.1	Data for public lands, national wild lands, scenic rivers, etc.
Recreation potential (SV2)	Presence of public parks or recreation areas	Occurrence Not occurrence	1 0	Data for federal, state, or city parks, etc.
Proximity to historic and cultural sites (SV3)	The number of historic and cultural sites	>3 ≥1 and <3 percentile None	1.0 0.5 0.1	Historic or cultural sites

the wetlands or related entities under assessment, such as the usage of wetlands (e.g., for recreation), hydrological facilities near wetlands (e.g., a dam), and riparian land use types and ownership.

18.2.2 DETERMINATION OF DECISION CRITERIA

Decision criteria are used to evaluate indicators based on the values of the corresponding metrics (Table 18.1). In our study, three methods were employed to determine the decision criteria based on (a) spatial statistics, (b) published results, or (c) professional judgments. The spatial statistics, such as the percentile, equal interval, standard deviation, or user-defined interval of metric values, were used to determine the decision criteria for ranking most of the selected indicators of wetlands. This statistical approach considers the variation of values of a particular indicator across the study area. Some decision criteria were adopted from the findings of published environmental studies or environmental management documents (e.g., ones relating to species density; USEPA 2002). As in many other instances of environmental decision making, professional judgments also played a role in determination of decision criteria for some indicators. Taking the indicator "Proximity to species of concerns (BV2)," for example, the decision criteria were adopted based on the published guidelines (USEPA 2002):

> If less than 2 species are found ("less vulnerable") near a wetland, the indicator receives a score of 0.1;
> If 2–5 species ("vulnerable"), the indicator receives a score of 0.5;
> If more than 5 species ("highly vulnerable"), the indicator receives a score of 1.0.

In addition, different weights were given to the species of concern based on their conservation status as endangered (a weight of 3), threatened (a weight of 2), or at risk (a weight of 1).

The decision criteria with the cutoff thresholds, like those above, may cause imprecise evaluation of metrics, especially when the metric value is close to the thresholds. For example, according to the thresholds of the decision criteria used for the above species indicator, a wetland that supports 3 species is ranked the same (a score of 0.5 or "vulnerable") with a wetland supporting 4 species. To take account of the vagueness and the nonspecificity of certain metrics values, a computational method based on fuzzy math was developed and applied to the evaluation of some indicators. With this method, the triangular-shaped fuzzy membership function (Tran et al. 2002) is utilized to determine the *degree of certainty (fuzzification)* of each metric value belonging to a certain vulnerability level, which is calculated by combining the portioned degree of certainty of the metric values in each vulnerability level. Taking the "Proximity to species of concerns (BV2)" indicator, for example, the fuzzification works as illustrated in Figure 18.2.

According to Figure 18.2, the wetland with 3 species of concern in its proximity has a "certainty" value of 0.25 in the "less vulnerable" domain, the value of 0.75 in the "vulnerable" domain, and the value of 0 for the "highly vulnerable" domain. For the wetland with 4 species of concern, the certainty values in these domains are 0, 1.0, and 0, respectively. When combining these values:

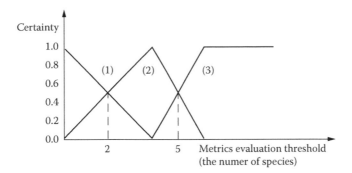

FIGURE 18.2   The triangular fuzzy membership functions for evaluating "species of concern" indicator. Membership function (1) denotes the "less vulnerable" level (the metric value = 0.1); (2) denotes the "vulnerable" level (the metric value = 0.5); (3) denotes the "highly vulnerable" level (the metric value = 1.0).

Metric value = 0.25 * 0.1 + 0.75 * 0.5 + 0.0 * 1.0 = 0.4 (for the 3 species case)

Metric value = 0.0 * 0.1 + 1.0 * 0.5 + 0.0 * 1.0 = 0.5 (for the 4 species case)

Thus, the fuzzy math–based method is more precise for distinguishing the difference between wetlands with 3 species and 4 species than using the cutoff thresholds that treats the two cases equally.

18.2.3 Calculation of Wetland Vulnerability Index

We developed calculations that incorporate the ranked scores of the indicators to generate both the vulnerability index for individual wetland functions and the overall vulnerability index for all wetland functions together. The vulnerability index for individual wetland functions is calculated using the following equation:

$$VI_k = W_1{}^*V_1 + W_1{}^*V_2 + ... + W_n{}^*V_n \qquad (18.1)$$

where:
VI_k = vulnerability index for the wetland function k
K = denoting an individual function: b for biological supporting function, h for hydrological function, p for physiographic function, and s for socioeconomic function, respectively
V_n = the ranked score value of the n*th* indicator of a particular function
W_n = the weight of n*th* indicator when it applies

Then, the vulnerability index for each wetland function (VI_k) is ranked in one of three possible vulnerability levels, "highly vulnerable," "vulnerable," or "less vulnerable," by evenly dividing the maximum value of VI_k for the assessment area into three intervals. To calculate the overall wetland vulnerability index, we first normalize each VI_k using the range of the score value of each wetland function:

$$NVI_k = (VI_k - VI_{min}) / (VI_{max} - VI_{min}) \qquad (18.2)$$

where:

NVI_k = normalized vulnerability index of the wetland function k,

VI_{min} = the sum of possible minimum score values of all indicators of the wetland function k,

VI_{max} = the sum of possible maximum score values of all indicators of the wetland function k.

The objective of the normalization is to treat all the functions equally in the index calculation by eliminating the effect of different score ranges among different wetland functions. Then the overall wetland vulnerability index (WVI) is calculated by combining the normalized vulnerability indices of all the wetland functions:

$$WVI = NVI_b + NVI_h + NVI_p + NVI_s \qquad (18.3)$$

The overall wetland vulnerability indices are ranked by evenly dividing the maximum value of WVI for the assessment area into three possible vulnerability levels: "highly vulnerable," "vulnerable," and "less vulnerable."

18.3 DECISION SUPPORT GIS FOR MODEL IMPLEMENTATION

Focusing on the implementation of the geospatial decision models for wetland vulnerability assessment, a decision support GIS (Figure 18.3) was developed with four major functions: (1) geospatial data management, (2) analytical query, (3) vulnerability assessment modeling, and (4) assessment result output. This was accomplished by customizing a widely used commercial GIS, ArcView (ESRI, Inc.), in order to fully utilize its capabilities in geospatial data handling, and increase the model's applicability and transferability in the community of users. Visual Basic of Application, an object-oriented language, was used to program ArcObject (the customizable components available with ArcView) in creation of the user-friendly graphical user interfaces (GUI) for implementing all the model functions.

18.3.1 Geospatial Data Management Function

A comprehensive wetland vulnerability assessment relies on efficiently retrieving sufficient data and information that address major concerns of wetland conservation. Therefore, a geospatial database management function is fundamental to the decision support GIS. This system function (Figure 18.4) is focused on two technical objectives that allow users to efficiently manage geospatial data for modeling:

1. Categorizing and organizing existing data sets. The interface shown in Figure 18.4a provides the user a tool for categorizing miscellaneous unorganized geospatial data into the classes that address major concerns in the assessment of biological, hydrologic, physiographic, and socioeconomic functions of a wetland.
2. Facilitating data retrieval in modeling. A dataset for addressing a particular concern or a group of data sets for addressing multiple concerns can be efficiently selected from corresponding data categories through the interface shown by Figure 18.4b.

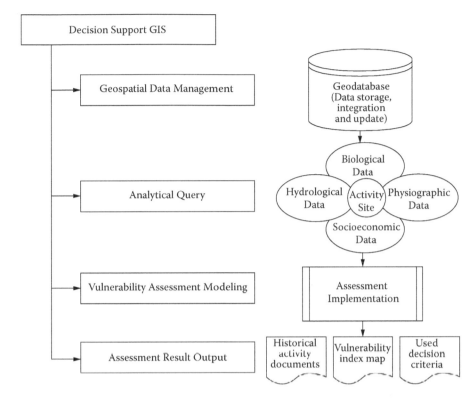

FIGURE 18.3   Key functions and architecture diagram of the decision support GIS described in section 18.3.

18.3.2 ANALYTICAL QUERY FUNCTION

As a GIS operation, the analytical query is referred to the retrieval, visualization, spatial analysis, or modeling of geospatial data in order to evaluate specific criteria or answer research questions quantitatively or qualitatively. A wetland vulnerability assessment usually requires the verification of spatial proximity or other relationships between the site of a proposed activity and a potentially impacted wetland or the locations of other entities of concern, such as historical permit sites, important habitats, biological resources, and cultural facilities. Therefore, the analytical query function of the decision support GIS was developed for the following major capabilities, which can be implemented through several customized interfaces (Figure 18.5):

1. Identifying and displaying the spatial location of a proposed activity in relation to a potentially affected wetland (Figure 18.5a). This is done by searching the proposed activity site with its known geographic coordinates or using the linguistic description (e.g., the name of a river) of the proposed activity location to identify the site on a background map.

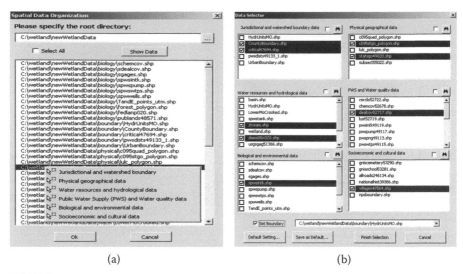

(a)   (b)

FIGURE 18.4 Two example interfaces designed for the geospatial database management function. Geospatial data are categorized into six classes for management in the modeling: (1) jurisdictional and watershed boundaries, (2) physical geographical data, (3) water resources and hydrological data, (4) public water supply and water quality data, (5) biological and environmental data, (6) socioeconomic and cultural data. (a) Categorizing miscellaneous, unorganized data into one of appropriate classes. (b) Facilitating geospatial data retrieval from organized data category folders.

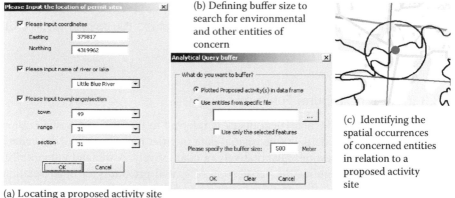

(a) Locating a proposed activity site

FIGURE 18.5 The illustration of user-friendly interfaces and capabilities for the analytical query function. (a) Locating a proposed activity site. (b) Defining buffer size to search for environmental and other entities of concern. (c) Identifying the spatial occurrences of concerned entities in relation to a proposed activity site.

2. Specifying spatial and temporal restrictions (e.g., distance, time, etc.), related attribute characteristics, or a combination of search criteria to identify concerned environmental or socioeconomic entities that might be affected by the proposed activity (Figure 18.5b). Graphical buffer zones (e.g., 500 meters) can be specified and generated around the proposed activity site or the wetland(s) to search the entities of concern within the buffer distance.

3. Identifying the spatial occurrences and measuring the spatial proximity of concerned environmental or socioeconomic entities in relation to the proposed activity site (Figure 18.5c). This will help answer specific analysis questions, such as "Are there any endangered species or important cultural facilities? How far are they from the proposed activity location?"

18.3.3 VULNERABILITY ASSESSMENT MODELING FUNCTION

When it comes to the implementation of geospatial decision models for wetland vulnerability assessment, a fundamental task is to evaluate the metrics of indicators based on geospatial data and analysis. A decision support GIS function was designed for this capability and includes spatial buffer generation, entity retrieval and visualization, spatial distance calculation, area measurement, and attribute information retrieval and verification. For the model implementation, a decision rule wizard (Figure 18.6a) was created to display all the indicators for each wetland function. Additionally, the assessment wizard (Figure 18.6b) was designed for all the necessary procedures ranging from indicator selection, metrics evaluation, decision criteria determination, weight selection, to fuzzy math–based evaluation of metrics.

18.3.4 ASSESSMENT RESULT OUTPUT FUNCTION

The assessment result output function includes the following capabilities:

1. Generating the vulnerability index maps for assessed wetlands using specially designed map symbols showing the levels of vulnerability. With these maps, the user can efficiently verify assessment results and make further decisions.
2. Storing the decision criteria used for assessed wetlands.
3. Managing historical activity documents for future reference.

18.4 MODEL APPLICATION

18.4.1 STUDY AREA AND GEOSPATIAL DATA

We applied developed decision models to the Little Blue River watershed, which is located on the fringe of Kansas City in the state of Missouri (United States) (Figure 18.7). The basin topography consists of rolling to hilly plains. The land cover of the watershed is mostly rural and dominated by cropland and grassland in addition to scattered forestland. Metropolitan Kansas City's suburbs are expanding eastward rapidly within the watershed, indicating the threats to the existing wetlands. Several perennial and intermittent rivers falling within the area have been experiencing substantial harmful human alterations. A large number of habitats have been converted to other land uses in order to enhance social benefits. Natural riverine processes, critical to providing ecosystem goods and services, have been greatly altered (U.S. National Research Council 2002). Flowing through the Kansas City metropolitan area, the Little Blue River is grossly polluted by point and nonpoint source pollutants due to extensive channel alterations. Degraded wetland habitats are thus obviously and continuously being modified in line with some human interests.

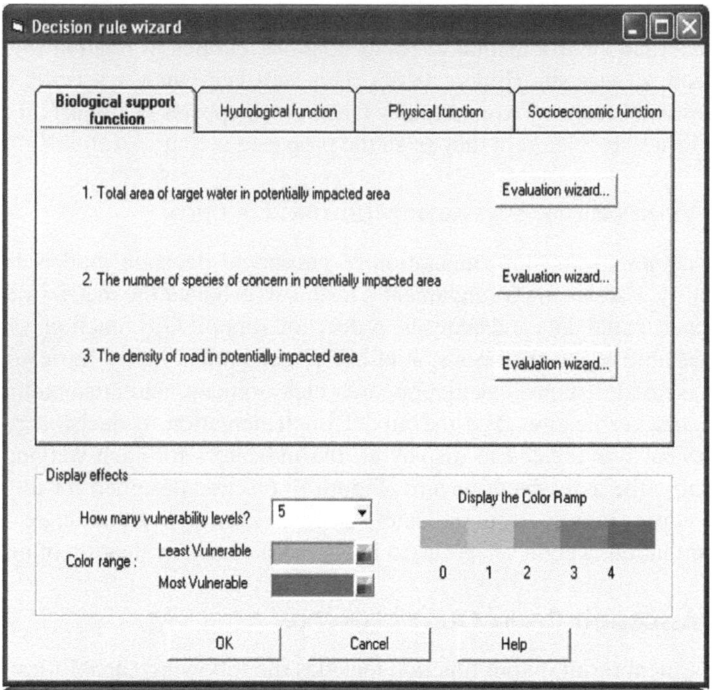

(a)

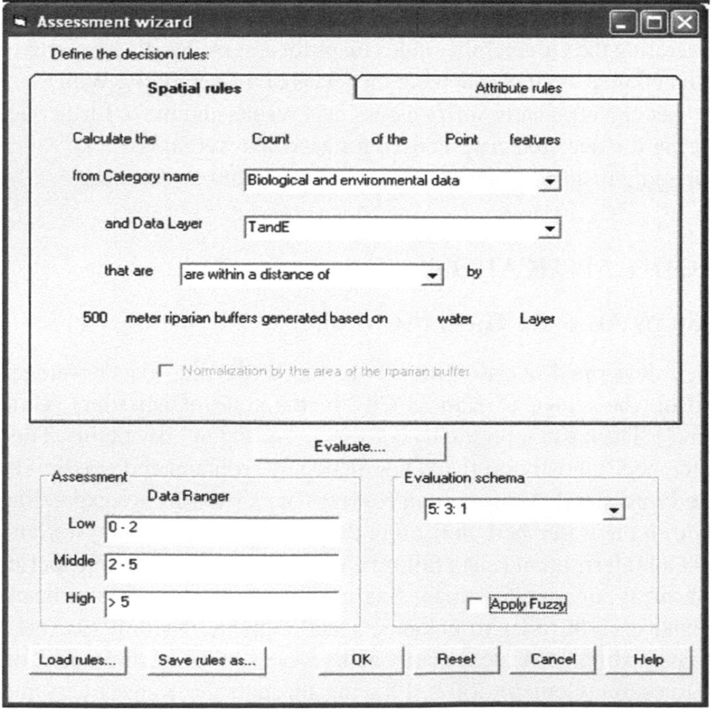

(b)

FIGURE 18.6 Illustrations of (a) the decision rule wizard and (b) the assessment wizard designed for implementation of geospatial decision models. (**See color insert after p. 162.**)

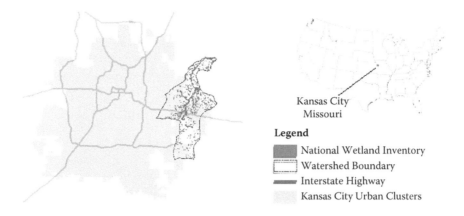

FIGURE 18.7   The assessment area in the Little Blue River watershed, Kansas City, Missouri. (**See color insert after p. 162.**)

In the assessment, we adopted a 500-meter riparian buffer zone as the potential impact area surrounding each wetland (Magee 1998). Our aim was to capture all the possible entities that could be impacted directly or indirectly by modifications of wetland regimes.

The boundaries of the Little Blue River watershed and its subwatersheds were delineated using the ArcHydro tool (Maidment and Djokic 2000) based on the 30-meter DEM (digital elevation model) data downloaded from the Web site of the Missouri Spatial Data Information Service (http://www.msdis.missouri.edu/). General occurrences of wetlands (Figure 18.7) were obtained from the National Wetland Inventory (NWI) maps at a scale of 1:24,000. To be compatible with the wetland definitions (1987) of the U.S. Army Corps of Engineers, preprocessing of the NWI data was conducted to filter out non-wetlands, such as farm ponds, permanently flooded areas, and other water regimes with drainages or other types of human activities. The wetland regime data were further refined based on visual interpretations using aerial photos (2003) obtained from the National Agricultural Inventory Program (NAIP). The refined wetland data, as shown in Figure 18.8 (left), include features of dominant (deciduous) forested wetlands located at water-accumulated spots near stream confluences.

18.4.2 Results and Validation

The map depicting overall vulnerability index values for all the wetlands in the Little Blue River watershed is presented in Figure 18.8 (middle). The land cover types were delineated by classifying satellite remote sensing data (Ji et al. 2006). The general spatial patterns were identified based on the results of the wetland vulnerability assessment. The wetland regimes located near urban centers had relatively high values of vulnerability. The wetlands that were more vulnerable are closer to the streams of higher order. The wetlands close to man-made lakes also demonstrated high vulnerability, probably caused by high water volumes impounding from dams. In the middle part of the watershed, several wetlands were highly vulnerable, as they are serving as natural buffers preventing residential areas downstream from

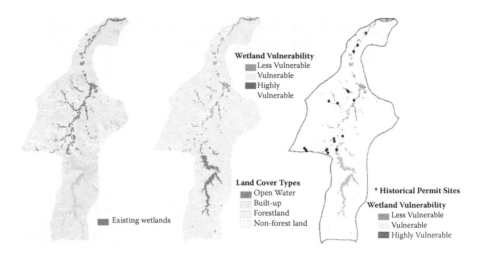

FIGURE 18.8 Vulnerability assessment results for the Little Blue River watershed. **(See color insert after p. 162.)**

being flooded during heavy rainfall. These wetlands are worthy of special attention and protection.

To validate the results of the assessment, we compared the regional wetland vulnerability map, which was generated using these models, with the locations of the human activities that were previously permitted through Section 404 reviews (data source: U.S. Army Corps of Engineers in Kansas City), assuming that the latter would usually occur at the wetland regimes of low vulnerability. (Note: the activities permitted under the mitigation bank option may be an exception to this assumption. In this case such activity locations should be excluded from use in the validation.) The permit location data show that between 1999 and 2000, there were 19 such permitted human activities near the existing wetland regimes in the Little Blue River watershed (Figure 18.8, right). The comparison indicated that none of the 19 sites were at highly vulnerable wetland regimes, with 14 sites at less vulnerable wetlands and 5 sites by vulnerable wetlands. This significant agreement between the two data sets validates the effectiveness of the geospatial decision models developed through this project.

18.5 CONCLUSIONS AND DISCUSSIONS

The test assessments in the Little Blue River watershed demonstrate the effectiveness of geospatial decision models in identification of vulnerable wetlands. The wetland vulnerability across the watershed reveals distinguishable spatial variation in relation to streams and land uses.

The model-generated wetland vulnerability indices and their distribution maps can be used as the prewarning information for regulatory wetland assessment, such as the CWA Section 404 permit review and other related wetland managements.

Together with the GIS-based decision support system, the geospatial decision models demonstrate the following technical advantages, which make the geospatial data–based wetland assessment an effective and efficient tool for management decision support:

1. Wetland vulnerability assessment can be carried out at any geographic scale within a watershed, allowing the linkage between landscape characteristics and wetland condition evaluation.
2. The indicators used in the models are designed such that they can be completely measured by analyzing geospatial data rather than through time-consuming filed surveys. As such, the decision criteria for evaluating wetland vulnerability can be dynamically determined based on the geospatial statistics of values of individual indicators across various landscape conditions in a given geographic area. This provides a frequency basis of land characteristics for reducing intrasite scale effects.
3. The fuzzy math–based algorithm can increase the precision of evaluations of indicator metrics. The customized decision support GIS provides user-friendly interfaces, which make it possible to rapidly assess wetlands with multiple criteria at changing geospatial scales.

REFERENCES

Adamus, P. R. 1983. *A method for wetland functional assessment.* FHWA-IP-82-23. Washington, DC: U.S. Department of Transportation, Federal Highway Administration.

Brinson, M. M. 1996. Assessing wetland functions using HGM. *National Wetlands Newsletter* 18(1):10–16.

Cook, R. A., A. J. Lindley Stone, and A. P. Ammann. 1993. *Method for the evaluation and inventory of vegetated tidal marshes in New Hampshire.* Concord: Audubon Society of New Hampshire.

Dodds, W. K. 2002. *Freshwater ecology. Concepts and environmental applications.* San Diego, CA: Academic Press.

Hollands, G. G., and D. W. Magee. 1985. A method for assessing the functions of wetlands. In *Proceedings of the national wetland assessment symposium*, ed. J. Kusler and P. Riexinger. Berne, NY: Association of Wetland Managers, 108–118.

Hruby, T., W. R. Cesaneck, and K. F. Miller. 1995. Estimating relative wetland values for regional planning. *Wetlands* 15(2):93–107.

Ji, W., J. Ma, R. A. Wahab, and K. Underhill. 2006. Characterizing urban sprawl using multistage remote sensing images and landscape metrics. *Computers, Environment and Urban Systems* 30:861–879.

Ji, W., and J. B. Johnston. 1994. A GIS-based decision support system for wetland permit analysis. *GIS/LIS'94 Proceedings.* Bethesda: ACSM-ASPRS-AAG-URISA-AM/FM, 1994. 1:470–475.

Ji, W., and J. B. Johnston. 1995. Coastal ecosystem decision support GIS: Functions and methodology. Marine & Coastal GIS issue (Ronxing Li, ed.), *Marine Geodesy*, 18(3):229–241.

Ji, W., and L. C. Mitchell. 1995. An analytical model-based decision support GIS for wetland resource management. In *Wetland and environmental applications of geographic information systems*, ed. John Lyon and Jack McCarthy. Boca Raton, FL: Lewis Publishers, 31–45.

Karr, J. R. 1997. Measuring biological integrity. Essay 14A. In *Principles of conservation biology*, 2nd ed., ed. G. K. Meffe and G. R. Carroll. Sunderland, MA: Sinauer, 483–485.

Karr, J. R., and E. W. Chu. 1998. *Restoring life in running waters: Better biological monitoring.* Covelo, CA: Island Press.

Magee, D. W. 1998. *A rapid procedure for assessing wetland functional capacity based on hydrogeomorphic classification.* Bedford, NH: Normandeau Associates (available from the Association of State Wetland Managers, Berne, NY).

Maidment, D., and D. Djokic. 2000. *Hydrologic and hydraulic modeling support: With geographic information systems.* Redlands, CA: ESRI Press.

Mitsch, W. J., and J. G. Gosselink. 2000. *Wetland,* 3rd ed. New York: John Wiley & Sons.

Sharitz, R. R., and D. P. Batzer. 1999. An introduction to freshwater wetlands in North American and their invertebrates. In *Invertebrates in freshwater wetlands of North America: Ecology and management,* ed. D. P. Batzer, R. B. Rader, and S. A. Wissinger. New York: Wiley, 1–22.

Sousa, P. J. 1985. *Habitat suitability index models: Red spotted newt.* U.S. Fish and Wildlife Service, Biological Report 82 (10.111). Washington, DC: U.S. Fish and Wildlife Service.

Stein, E. D. 1998. A rapid impact assessment method for use in a regulatory context. *Wetlands* 18(3):379–392.

Tran, L. T., C. G. Knight, B. V. O'Neill, E. R. Smith, K. H. Riitters, and J. Wickham. 2002. Fuzzy decision analysis for integrated environmental vulnerability assessment of the mid-Atlantic region. *Environmental Management* 29(6):845–859.

U.S. Army Corps of Engineers (USACOE). 1997. National action plan to implement the hydrogeomorphic approach to assessing wetland functions. *Federal Register* 62(119):33607–33620.

U.S. Environmental Protection Agency (USEPA). 2002. Index of watershed indicators: An overview. Office of Wetlands, Oceans, and Watersheds. 38.

U.S. Fish and Wildlife Service (USFWS). 1980. *Habitat evaluation procedure (HEP) manual* (102 ESM). Washington, DC: U.S. Fish and Wildlife Service.

U.S. National Research Council. 2002. Committee on Missouri River Ecosystem Science. *The Missouri River ecosystem: Exploring the prospects for recovery.* National Academy Press, Washington, D.C.

Zampella, R. A., et al. 1994. *A watershed-based wetland assessment method for the New Jersey pinelands.* New Lisbon, NJ: Pinelands Commission.

19 Watershed Science
Essential, Complex, Multidisciplinary, and Collaborative

*R. Jan Stevenson, Michael J. Wiley, Stuart H. Gage,
Vanessa L. Lougheed, Catherine M. Riseng,
Pearl Bonnell, Thomas M. Burton,
R. Anton Hough, David W. Hyndman,
John K. Koches, David T. Long,
Bryan C. Pijanowski, Jiaquo Qi, Alan D. Steinman,
and Donald G. Uzarski*

19.1 WATERSHED SCIENCE: ESSENTIAL

Sustainability of ecosystem services for human well-being will require thinking at multiple spatial and temporal scales (Kates et al. 2001). Large-scale assessment of global change provides an overview of the diversity of environmental problems that are occurring and are likely to occur in the future (Millennium Ecosystem Assessment 2005). However, even global assessments require scaling to smaller areas to account for local variations in ecosystems, human activities affecting those ecosystems, and societal values that value different elements of ecosystems. Watersheds provide an important geospatial unit for the science of water resource management because of the greater interaction between humans and ecosystems within watershed boundaries than across watershed boundaries.

In this paper, we describe the importance of watershed science for watershed management and regional sustainability. We also describe how scientists and other stakeholders from many disciplines must work together to solve and prevent environmental problems, and that those collaborations have great benefits for the individuals involved, their science, and society. To illustrate the concepts discussed, examples will be provided from the Muskegon Watershed Research Partnership (MRWP). The MRWP conducts an integrated research effort on one of the largest and most ecologically diverse watersheds of the Great Lakes region in the United States.

Scientists and policy makers often lament the lack of research and knowledge that are necessary to make the difficult decisions that frequently face resource managers (Brewer and Stern 2005). Scientists need to provide information that answers the following fundamental questions: Is there a problem in our ecosystem? What

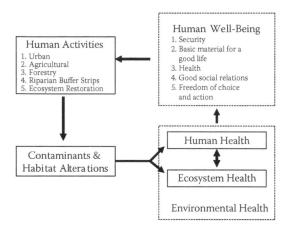

FIGURE 19.1 Conceptual figure of the role of watershed management in support of human health, ecological health, and thereby human well-being.

is causing the problem? How can we fix the problem? These problems range from large-scale climate change problems to many regional problems related to land use and water quality in watersheds. Research has obviously provided knowledge to help identify problems and solutions, but often it does not go far enough, such as characterizing the frequency, intensity, and duration of problems, their direct linkage to human well-being, and the risk of problems with specific levels of contamination and habitat alteration.

In this paper, we work at the watershed scale to consider ecological health questions related to alterations and contamination of water bodies by human activities. However, we work within a larger conceptual framework than traditional water quality assessment (Figure 19.1) because it considers the feedback of human activities on all aspects of human well-being. Human activities produce contaminants and habitat alterations that affect both ecological and human health. Ecological and human health, as well as many other factors, are measurable attributes of ecosystem services and human well-being (Parris and Kates 2003). In this model, human and ecological systems are integrated, but categorized into four groups by the way information is used in environmental management. The delineation and numbers of categories are not as important as identifying the interrelationships of elements within these categories and how they are used. Such information can be used to manage human activities with geospatially aware policies that minimize contaminants or habitat alterations and optimize ecological and human health. Tradeoffs among elements of human well-being are important considerations for management actions. Based on management actions, human activities can be altered and regulated to minimize contaminants and habitat alterations.

Sound science is essential for characterizing environmental condition and determining whether or not problems exist, for diagnosing the contaminants and habitat alterations that are causing problems, and for developing management options for solving these problems. To assess problems, we need to be able to precisely monitor valued ecological attributes (VEAs) such as biodiversity, fisheries production, and water quality for human health. In addition, the technical expertise is needed for accurate measurement of contaminants and habitat alterations to determine whether or not they could be causing problems. Knowledge of relationships among VEAs and contaminants or habitat alterations, and how to detect these relationships, is necessary to characterize risks of problems at specific VEA levels. Indeed, knowledge of relationships among human activities, contaminants and habitat alterations, VEAs, human health, and measures of human well-being will enable more rigorous evaluation of the tradeoffs that may be necessary between short-term economic growth and long-term sustainable environmental quality (Kates et al. 2001).

19.2 WATERSHED SCIENCE: COMPLEX

Complexity in watershed science is caused by many factors, such as: the many VEAs for which watersheds are managed; the many human activities, contaminants, and habitat alterations that affect VEAs and human well-being; nonlinearity in simple relationships and synergistic or antagonistic interactions in more complex relationships; new processes being important as spatial scales expand; and time lags in cause–effect interactions. This complexity can be managed to provide the information needed to make management decisions with reasonable certainty of success. Organizing the information gathering, analysis, and decision-making processes helps solve complex problems.

Frameworks for environmental assessment and management can be used to organize the problem-solving process and to list the issues that should be considered. Numerous frameworks have been proposed and, fortunately, they all have much in common (e.g., USEPA [U.S. Environmental Protection Agency] 1996, 2000b, Stevenson et al. 2004a). They all emphasize the continual process of assessment and management to refine management strategies and to ensure and improve management results. In general, frameworks developed by natural scientists start with defining the problem in terms of ecological or human health, whereas political scientists would argue that we should start by considering what is needed to make a good management decision (Dietz 2003). The latter approach has merit because it sets the breadth of scope of factors that should be considered in an environmental assessment, which should include economic and social factors as well as ecological and human health goals.

Because the MWRP assessment is based on the framework of Stevenson et al. (2004a), it incorporates concepts from the broader work of Barbour et al. (2004), Suter (1993), and the USEPA (1996). The framework in Stevenson et al. (2004a) emphasizes four stages: study design, environmental characterizations, stressor diagnosis, and management (Figure 19.2).

19.2.1 ASSESSMENT DESIGN

The first MWRP effort involved gathering stakeholders from the watershed to list the environmental issues that should be considered, and then developing a series of plans for implementing the assessment. This was the very beginning of the design stage where objectives of the assessment were determined. The stakeholders included representatives from environmental groups, businesses that would be regulated, local

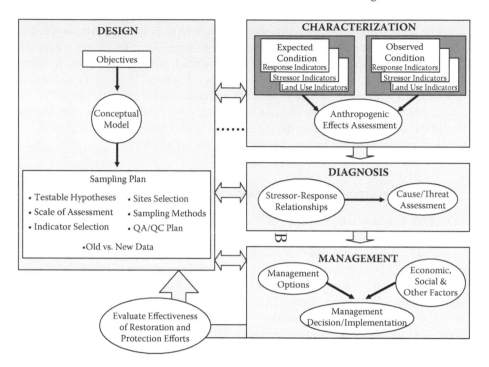

FIGURE 19.2 The protocols for ecological assessment related in a framework. Three major steps, study design, analysis, and integration, are emphasized (from Stevenson et al. 2004a).

and state government agencies, the scientific community, and funding agencies that would support the ongoing work of the MRWP. A series of meetings was held to discuss the range of issues that should be considered and how the assessment should be done. Stakeholders then organized themselves into working groups so that each group was involved with a slightly different but interrelated project that would contribute to the watershed assessment.

In the design stage, the MRWP defined objectives, developed a conceptual model linking human activities to environmental conditions, and developed a sampling plan. Among the many ecosystem services supported by the Muskegon River watershed (MRW), stakeholders decided that biological condition and fisheries production were valued ecological attributes that should be "endpoints" in our assessment. Stakeholders also decided that conditions of streams, lakes, and wetlands should be studied.

Biological condition and fisheries production are two "uses" for which the state of Michigan manages its waters. Uses have significance in U.S. regulations because they define the goals of environmental management (USEPA 1994). Management actions in the United States are triggered by violations of environmental criteria, which are directly related to supporting specific uses of waters. For example, the use of a water body may be defined as warm-water aquatic life use, as well as recreation, domestic or industrial water supply, and navigation. All waters are assigned to support from one to all uses. When a use is assigned for a water body, then either narrative or numerical characterizations of conditions of that water body are established as criteria that indicate support of that use. Narrative criteria might be "absence of nuisance algal growths" or "natural balance of flora and fauna." Numeric criteria relate to specific, quantitative levels of species composition, species diversity, and productivity of the habitat. Uses also have value as targets of environmental research because we need to understand how human activities affect contaminants and habitat alterations, and how those directly affect uses. Environmental research can help quantify this understanding sufficiently to justify numeric criteria, which reduces the ambiguity of interpreting narrative criteria while making management decisions.

Fisheries production in the Great Lakes region is commercially important for the sport-fishing industry, which is a major recreational industry in the Great Lakes region. Biological condition is an important attribute of ecological assessment because it provides a good indicator of structure and function of ecosystems (Angermeier and Karr 1994), and thereby supports an integrated assessment of the four types of ecosystem services described in the Millennium Ecosystem Assessment (2005): supporting, provisioning, regulating, and cultural. The U.S. Clean Water Act calls for "protecting the physical, chemical, and biological integrity." Biological integrity is a high level of biological condition. Biological condition can be measured as deviation from a natural or some other desired condition (Hughes 1995). Biological condition is a relatively precise indicator compared to some temporally variable measures of physical and chemical condition because the biota present in a habitat reflect historic as well as current physical and chemical conditions in the habitat. Waters with high biological integrity are assumed to be safe for many other uses because waters are close to natural. Thus, biological condition has been adopted as an essential element in water quality assessments under the U.S. Clean Water Act and the Water Framework Directive of the European Union.

After identifying overall targets for the assessment, the MRWP developed a study design to interrelate all elements of the assessment. The MRWP project had three basic objectives. First, we wanted to assess the current condition of streams, lakes, and wetlands in the watershed. Second, we wanted to develop a model of the system so that we could predict results of management actions and forecast future changes in the watershed under different management scenarios. Finally, we wanted to communicate results of our assessment to the stakeholders, including the public and government officials.

Stakeholders then developed a conceptual model of the system that we wanted to understand. With stakeholder delineation of biological condition and fisheries production as VEAs for the project, we needed to determine the other factors that were likely important in the relationships among human activities, contaminants and habitat alterations, and VEAs. Based on previous knowledge, the MRWP hypothesized that dams, sediments, habitat loss, stream channelization, nutrient enrichment, and invasive species would be important contaminants and habitat alterations affecting biological condition and fisheries production. Human health endpoints associated with microbial contamination of recreational areas were assessed as important, but were considered beyond the scope of studies that could be afforded at the time. It was also decided that land use and land cover would be important factors to consider in the assessment and modeling that would need to distinguish between natural variation in ecosystems versus human effects on ecosystems. Finally, land use in the past as well as the future was considered to account for legacies of past activities in the watershed.

To achieve the three MRWP objectives, several project modules emerged, each with a team of scientists and stakeholders having expertise and interest in accomplishing tasks in the project defined by the scope of the conceptual model. All projects involved with the assessment and modeling adopted the general conceptual model illustrated in Figure 19.1, in which human activities, contaminants and habitat alterations, and VEAs were specifically related. One project focused on refining land use land cover characterizations of the watershed. Another project had the responsibility of assessing biological condition, fisheries production, and contaminants and habitat alterations in the watershed. A third major project was responsible for synthesizing results of the assessment and developing an integrated, process-based watershed-scale model. Other projects on economic development, human health, and methods for communicating results to the public received less funding or were postponed until funding opportunities develop. As a result, all projects that were funded assumed responsibility for communicating results to stakeholders.

Land use land cover characterizations were designed to characterize natural features and human activities in the watershed. Both natural and human features of the landscape are important for characterizing the natural potential for a water body, how human activities have affected it, and how human activities can be regulated to minimize effects. Satellite imagery was used to characterize land use land cover in the watershed. Extensive ground truthing was conducted by field crews. This information was made available to the teams working on water body assessment and the watershed model.

Assessment of biological condition, fisheries production, and contaminants and habitat alterations in water bodies of the MRW involved developing a detailed sampling plan that would achieve the objectives of characterizing conditions and diagnosing causes and threats to VEAs. We used three different approaches for sample site selection to achieve three slightly different objectives in our assessment.

1. To characterize the condition of all water bodies in the watershed, we selected sampling sites within the watershed using a random sampling design stratified by water body type (streams, lakes, and wetlands). Random sampling enables scaling assessments from a fraction of all water bodies to an unbiased estimate of conditions in all streams, lakes, and wetlands in the watershed.
2. A stratified random sampling design with strata defined by water body type and land use was used to develop stressor-response relationships between VEAs, contaminants and habitat alterations, and human activities. Stressor-response relationships were going to be important for diagnosing causes and threats of VEA impairment (Figure 19.2) and in refinements of more complex, process-based watershed models. This called for sampling outside the boundary of the MRW to find sufficient numbers of water bodies with higher levels of human activity.
3. Sites also were selected because of special interest by stakeholders. For example, all large lakes in the watershed were selected because of their economic importance. Intensive sampling was also targeted in the lower Muskegon River where the Great Lakes sport fishing is concentrated.

The variables that we selected for measurement varied among streams, lakes, and wetlands. The same land use land cover variables were selected for each water body type. Similar chemical variables were measured in all habitats, except for more detailed trace-element studies in rivers. The latter substudy was designed to use ratios among trace elements to provide landscape signatures of human activities; these signatures are being used as another line of evidence of the relative importance of different levels of human activities in watersheds (Wayland et al. 2003). Different physical variables were measured in each water body type due to the nature of their physical structures.

Multiple biological attributes were measured for each water body type to provide more thorough assessments from the perspective of differing responses to stressors and to increase precision of water body assessments with multiple lines of evidence and multiple measurements. Biological attributes measured in each water body type varied depending upon the diversity of biological assemblages in that water body type and the likelihood of developing precise metrics of biological condition with the assemblages. Algae and benthic macroinvertebrates were measured in each water body type. Planktonic algae were assessed in lakes and benthic algae were assessed in streams and wetlands. Meiofauna such as zooplankton were assessed in wetlands and lakes. Fish were assessed primarily in streams and rivers. Fish data collected as part of government studies will be used for lake assessments.

New indicator development was an important project of the MRWP. New modifications of biological metrics will be made to improve their application for the MRW and for application in streams, lakes, and wetlands. In addition, new variables are being assessed in the MRW. For example, remote sensing methods are being refined to more accurately assess algal biomass in lakes and vegetation type and productivity in wetlands. Sound variables are also being used to characterize the level of human activity in watersheds and the biological condition of birds and amphibians that can be heard.

19.2.2 Assessment Characterization

Characterizing condition requires comparison of expected and observed conditions in VEAs and both contaminants and habitat alterations (Figure 19.2). Land use land cover measurements are important for defining expected condition and developing tools to diagnose problems for and threats to VEAs. Expected condition can be defined in many ways (Stevenson et al. 2004a): a desired condition such as high fisheries production; an a priori legally defined standard; the natural condition occurring if human effects were very low; or some acceptable deviation from natural conditions. Characterizing condition in an environmental assessment is then defined as the deviation in observed condition at a site from the expected condition for that site.

Expected condition in many assessment programs is based on the concept of reference condition (Hughes et al. 1986). Reference conditions can be characterized as (1) minimally disturbed in the region, (2) the best attainable with restoration, or (3) natural (Stoddard et al. in press). Extensive literature covers characterization of reference condition (e.g., Hughes and Larsen 1988, Hughes 1995, Barbour et al. 1999, Hughes et al. 2000). We chose two methods for defining expected condition:

1. A reference condition for minimally disturbed sites in the region will be the 75th percentile of the frequency distribution of attributes at sites that have low levels of human disturbance in watersheds. This approach is commonly used for ecological assessments (European Commission 2000, Hughes et al. 2000).

2. A regression-based method for defining expected condition based on natural conditions (with human disturbance close to zero) and variations in natural conditions, which was proposed by Wiley et al. (2002). Thus, expected condition varies among habitat types and is refined for natural variability among sites of the same water body type. For example, low-gradient, warm-water streams will have a different expected condition than high-gradient, cold-water streams. Large, deep lakes will have different expectations than small, shallow lakes. The advantages of Wiley's method include a more standardized comparison of observed condition to a natural reference condition, a refinement of characterizations based on natural variation among sites, and the ability to develop these predictive models of expected condition when few high-quality sites exist.

19.2.3 Assessment Diagnosis

Toxicological literature (Suter 1993, Lippman and Schlesinger 2000, USEPA 2000a) has reviewed numerous methods for diagnosing the contaminants and habitat alterations that pose the greatest threats to VEAs or are the likely causes of problems with VEAs. Stressor-response relationships, in this case between VEAs and contaminants or habitat alterations, are essential for relating observed conditions in habitats to likely risks of impairment due to specific contaminants and habitat alterations (Stevenson et al. 2004b). Although deviation of physical, chemical, and non-native species characteristics at a site from the expected condition for that site can be used to list potential causes of impairment, diagnosis of the contaminants and habitat alterations that most likely threaten or cause impairment of VEAs is more certain with quantitative stressor-response relationships. For example, changes in some physical and chemical attributes may have little effect on VEAs, whereas others have great effects.

Stressor-response relationships can be developed with experimental and field-survey results (Figure 19.3). Laboratory bioassays and even field experiments can be used to determine stressor-response relationships where levels of contaminants and habitat alterations are experimentally manipulated. While experimental approaches such as these are extremely valuable for documenting cause-effect relationships, transferal of results to large-scale field situations may be problematic. Experiments, by their nature, are typically conducted at much shorter temporal scales and smaller spatial scales than long-term, large-scale responses of ecological systems to contaminants and habitat alterations. Thus, stressor-response relationships based on field data are particularly valuable for determining the levels of contaminants and habitat alterations that cause unacceptable changes in VEAs.

In the MRW assessment, thresholds in stressor-response relationships will be used to establish benchmarks for contaminants and habitat alteration that cause

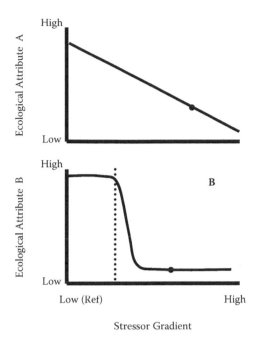

FIGURE 19.3 Relationships between valued ecological attributes and contaminants are important for both protection and restoration of ecosystems. In this figure, attribute A responds linearly to a contaminant or habitat alteration (stressor), whereas attribute B shows assimilative capacity of the stressor until a threshold response is observed. Threshold responses help justify benchmark stressor levels, which can be used as targets for restoring habitats or levels of protection for high-quality habitats. Attributes with threshold responses help justify designation of specific levels of stressors as management targets (Muradian 2001, Stevenson et al. 2004b).

unacceptable changes in VEAs (Stevenson et al. 2002, King and Richardson 2003). Thresholds are delineated as sudden changes in VEAs with relatively small increases in stressors (Figure 19.3). Benchmarks for contaminants and habitat alterations that cause threshold responses will be used to calculate *hazard quotients* (Suter 1993, Tannenbaum et al. 2003), which are ratios of observed conditions to the benchmark condition. Hazard quotients are also referred to as *toxic units* in some toxicological literature. Higher hazard quotients indicate a higher likelihood that a contaminant or habitat alteration is either threatening or causing impairment of VEAs. If nonlinear relationships between stressors and VEAs are not observed, then alternative methods will be used to establish stressor benchmarks (e.g., Setzer and Kimmel 2003).

Changes in sensitive and tolerant species will also be used to diagnose the contaminants and habitat alterations causing impairment. Relative sensitivities and tolerances of many organisms to pollution are documented in the literature. By comparing changes in species composition of observed sites compared to reference sites, inferences can be made about likely stressors. This provides another line of evidence to support diagnoses with hazard quotients.

19.2.4 ASSESSMENT FORECASTING

Although forecasting environmental change is not usually part of watershed assessments, it has become important in climate change assessments and in large-scale assessments of global environmental change. The MRWP includes assessment forecasting using traditional as well as innovative approaches. The more traditional methods of assessment forecasting may predict the effects of managing human activities in the watershed when human activities are regulated in different ways, such as dam removals, reduced fertilizer application and groundwater withdrawal, bank stabilization to reduce sedimentation, and restoration of riparian buffer zones. Responses to these management actions can be predicted with a system of process-based models. This is part of the responsibility of the MRWP modeling project, which is developing an integrated modeling system that includes watershed and in-stream hydrologic models, biogeochemical models, and biological response models (Wiley et al. this book).

Long-term forecasting of land use land cover change in a watershed and ecological results of future human activities can also be part of the assessment process. Forecasting assesses likely conditions in the future given current conditions and predicted changes to those conditions. The MRWP includes a special subproject that is developing a refined model of land use land cover change that is specifically calibrated to the MRW. This model is being integrated within the modeling system to predict long-term responses to different management actions (Tang et al. 2005).

19.2.5 MANAGING COMPLEXITY

The MRWP employs several methods to manage the complexity of the ecological systems being studied and of the assessment process itself. Focusing on clearly defined endpoints or VEAs helps limit factors being assessed to those related to the VEAs, in this case, biological condition and fisheries production. Use of an integrated assessment framework provided a roadmap of the steps to take and issues to consider during the assessment. This roadmap also included a list of tasks to be completed with assignments of responsibilities to each principle investigator (PI). Creating independent projects within the MRWP with optimal interconnection among projects also made the MRWP process more manageable and more likely to succeed. With tasks clearly identified for each project, the specific data required for the MRWP was also defined. The independent projects enable high levels of interaction within smaller groups of scientists without the cost of including all MRWP scientists in every conversation. As with natural systems, modularity also makes project systems more stable and easier to manage.

19.3 WATERSHED SCIENCE: MULTIDISCIPLINARY AND COLLABORATIVE

An extraordinary breadth of expertise is needed to assess and manage environmental problems. In the MRWP, which is limited to just the assessment, scientists from five universities and many more disciplines are involved. Most are natural scientists because the MRWP focuses on ecological health, but even this group is diverse and represents zoology, botany, geochemistry, entomology, limnology, hydrology, and geography. Without the experience of experts in each of these areas, our ability to conduct an integrated assessment at this scale would not be possible.

The success of the MRWP will be a result of it being multidisciplinary and collaborative. Interdisciplinary indicates interactions among disciplines, but higher levels of collaboration are required to integrate work thoroughly across so many disciplines. Collaboration is also required for brainstorming ideas about assessment approaches, variables to measure, methods of measurement, and methods of analysis. Compromises are necessary for this collaboration with all PIs sacrificing some of their independence to cooperate in the planning and timing of sampling and sample analysis. Collaboration will be an essential part of the interpretation and synthesis of our results as well as the planning and implementation of the project.

Rules of collaboration have been important for defining expectations of PIs, data sharing, and coauthorship on research papers. Establishment and agreement to these rules by PIs provided all with a sense of security for intellectual property. For the MWRP, we have rules for both sharing data and coauthorship of papers (see appendix A). Sharing data is critical for the multidisciplinary questions being asked, such as relating land use, hydrology, biogeochemistry, and biological condition. With data sharing come acknowledgments of contributions made by coauthors when results are shared during publication.

Collaboration has costs, particularly with time. Most importantly, good multidisciplinary collaborations with high levels of innovation require significant levels of communication among PIs. Extra time is needed for brainstorming ideas, making decisions, discussing and explaining plans, and coordinating the implementation of the project. This communication often takes extra time as colleagues from different disciplines learn the language of other disciplines. Many times a word has a different use or meaning in different disciplines. Specific tasks for specific teams within the project need to be detailed so they are conducted in a way that meets the objectives of the project. Those tasks need to produce data in formats that are compatible with a projectwide database and that have common features, such as data table format, codes for sites, and codes and units for variables. Detailing specifics before data collection is best, and that does not preclude modifying those plans when unforeseen problems develop.

Collaborations in multidisciplinary research provide great rewards that outweigh the sacrifices. First, all collaborators learn a great deal from their colleagues. New ideas and perspectives from different disciplines are particularly valuable and move all disciplines forward. This is particularly evident during the design and planning stages, but also during review of results. Getting input from colleagues across disciplines produces a more thorough and high-quality final product in the publications resulting from the research. Multidisciplinary projects often generate a core of information that can be leveraged against new projects. For example, the core data from the Muskegon assessment project has helped garner two to three times as much funding on complementary projects that build on this core dataset. And, finally, as noted in the introduction, close collaboration by an integrated multidisciplinary team of experts is essential for developing successful solutions for environmental problems.

ACKNOWLEDGMENTS

This MWRP research is funded by the Great Lakes Fisheries Trust and the Wege Foundation. Peter Wege, Bill Cooper, and Jack Bails led the effort to coordinate funding of the MWRP and the gathering of stakeholders to start the partnership.

REFERENCES

Angermeier, P. L., and J. R. Karr. 1994. Biological integrity versus biological diversity as policy directives. *BioScience* 44:690–697.

Barbour, M., S. Norton, R. Preston, and K. Thornton. 2004. *Ecological assessment of aquatic resources: Linking science to decision-making.* Pensacola, FL: Society of Environmental Toxicology and Contamination Publication.

Barbour, M. T., J. Gerritsen, and B. D. Snyder. 1999. *Rapid bioassessment protocols for use in wadeable streams and rivers,* 2nd ed. EPA 841-B-99-002. Washington, DC: United States Environmental Protection Agency.

Brewer, G. D., and P. C. Stern, eds. 2005. *Decision making for the environment: Social and behavioral research priorities.* Washington, DC: National Academies Press.

Dietz, T. 2003. What is a good decision? Criteria for environmental decision making. *Human Ecology Review* 10:60–67.

European Commission. 2000. Directive 2000/EC of the European Parliament and of the Council—Establishing a framework for community action in the field of water policy. Brussels, Belgium: European Commission.

Hughes, R. M. 1995. Defining acceptable biological status by comparing with reference conditions. In *Biological assessment and criteria: Tools for water resource planning and decision making,* ed. W. S. Davis and T. P. Simon. Boca Raton, FL: Lewis Publishers, 31–47.

Hughes, R. M., D. M. Larsen, and J. M. Omernik. 1986. Regional reference sites: A method for assessing stream potentials. *Environmental Management* 10:629–635.

Hughes, R. M., and D. P. Larsen. 1988. Ecoregions: An approach to surface water protection. *Journal of the Water Pollution Control Federation* 60:486–493.

Hughes, R. M., S. G. Paulsen, and J. L. Stoddard. 2000. EMAP—surface waters: A multiassemblage, probability survey of ecological integrity in the USA. *Hydrobiologia* 422:429–443.

Kates, R. W., W. C. Clark, R. Corell, J. M. Hall, C. C. Jaeger, I. Lowe, J. J. McCarthy, H. J. Schellnhuber, B. Bolin, N. M. Dickson, S. Faucheux, G. C. Gallopin, A. Grübler, B. Huntley, J. I. Jäger, N. S. Jodha, R. E. Kasperson, A. Mabogunje, P. Matson, H. Mooney, B. Moore III, T. O'Riordan, and U. Svedlin. 2001. Sustainability science. *Science* 292:641–642.

King, R. S., and C. J. Richardson. 2003. Integrating bioassessment and ecological risk assessment: An approach to develop numeric water quality criteria. *Environmental Management* 31:795–809.

Lippman, M., and R. B. Schlesinger. 2000. Toxicological bases for the setting of health-related air pollution standards. *Annual Review of Public Health* 21:309–333.

Millennium Ecosystem Assessment. 2005. *Ecosystems and human well-being: Synthesis.* Washington, DC: Island Press.

Muradian, R. 2001. Ecological thresholds: A survey. *Ecological Economics* 38:7–24.

Parris, T. M., and R. W. Kates. 2003. Characterizing and measuring sustainable development. *Annual Review of Environment and Resources* 28:559–586.

Setzer, R. W., Jr., and C. A. Kimmel. 2003. Use of NOAEL, benchmark dose, and other models for human risk assessment of hormonally active substances. *Pure Applied Chemistry* 75:2151–2158.

Stevenson, R. J., R. C. Bailey, M. C. Harass, C. P. Hawkins, J. Alba-Tercedor, C. Couch, S. Dyer, F. A. Fulk, J. M. Harrington, C. T. Hunsaker, and R. K. Johnson. 2004a. Designing data collection for ecological assessments. In *Ecological assessment of aquatic resources: Linking science to decision-making*, ed. M. T. Barbour, S. B. Norton, H. R. Preston, and K. W. Thornton. Pensacola, FL: Society of Environmental Toxicology and Contamination, 55–84.

Stevenson, R. J., R. C. Bailey, M. C. Harass, C. P. Hawkins, J. Alba-Tercedor, C. Couch, S. Dyer, F. A. Fulk, J. M. Harrington, C. T. Hunsaker, and R. K. Johnson. 2004b. Interpreting results of ecological assessments. In *Ecological assessment of aquatic resources: Linking science to decision-making*, ed. M. T. Barbour, S. B. Norton, H. R. Preston, and K. W. Thornton. Pensacola, FL: Society of Environmental Toxicology and Contamination, 85–111.

Stevenson, R. J., Y. Pan, and P. Vaithiyanathan. 2002. Ecological assessment and indicator development in wetlands: The case of algae in the Everglades, USA. *Verhandlungen Internationale Vereinigung für Theoretische und Andgewandte Limnologie* 28:1248–1252.

Stoddard, J. L., D. P. Larsen, C. P. Hawkins, R. K. Johnson, and R. H. Norris. In press. Setting expectations for the ecological condition of streams: The concept of reference condition. *Ecological Applications*.

Suter, G. W. 1993. *Ecological risk assessment*. Boca Raton, FL: Lewis Publishers.

Tang, Z., B. A. Engel, B. C. Pijanowski, and K. J. Lim. 2005. Forecasting land use change and its environmental impact at a watershed scale. *Journal of Environmental Management* 76:35–45.

Tannenbaum, L. V., M. S. Johnson, and M. Bazar. 2003. Application of the hazard quotient method in remedial decisions: A comparison of human and ecological risk assessments. *Human and Ecological Risk Assessment* 9:387–401.

U.S. Environmental Protection Agency (USEPA). 1994. Water quality standards handbook, 2nd ed. EPA/823/b/94/005a, Washington, DC: U.S. Environmental Protection Agency.

U.S. Environmental Protection Agency (USEPA). 1996. Strategic plan for the Office of Research and Development. EPA/600/R/96/059. Washington, DC: U.S. Environmental Protection Agency.

U.S. Environmental Protection Agency (USEPA). 2000a. Stressor identification guidance document. EPA 600/R/96/059. Washington, DC: U.S. Environmental Protection Agency.

U.S. Environmental Protection Agency (USEPA). 2000b. Toward integrated environmental decision-making. EPA/SAB/EC/00/001. Washington, DC: U.S. Environmental Protection Agency.

Wayland, K. G., D. T. Long, D. W. Hyndman, B. C. Pijanowski, S. M. Woodhams, and S. K. Haack. 2003. Identifying relationships between baseflow geochemistry and land use with synoptic sampling and R-mode factor analysis. *Journal of Environmental Quality* 32:180–190.

Wiley, M. J., P. W. Seelbach, K. Wehrly, and J. Martin. 2002. Regional ecological normalization using linear models: A meta-method for scaling stream assessment indicators. In *Biological response signatures: Indicator patterns using aquatic communities*, ed. T. P. Simon. Boca Raton, FL: CRC Press, 197–218.

Appendix

MUSKEGON PARTNERSHIP DATA SHARING AGREEMENT

Because of the collaborative nature of these projects, timely production and sharing of data is critical to everyone's success. Researchers participating in partnership projects can share data via the GVSU (Grand Valley State University) server or by directly contacting other researchers. Participating PIs are expected to both contribute relevant data themselves, and to use data from other collaborators in a responsible fashion. Data set expectations should be negotiated for each specific proposal task. The designated Task Leader and the funding Project Lead will be responsible together for developing data set expectations for specific proposal tasks and for publishing them in a Project Data Catalog on the public web site.

Because our goal is good science, collaborators (as a condition of continued funding) agree to a set of basic data-sharing principles and ethics as a condition of access. Details vary depending on the designated *release status* of particular data set; of which there are currently three:

Type 1: Public Domain: These are data available through the Web site or FTP server which are either in the public domain as a matter of law (e.g. certain monitoring data) or have been released for public distribution by collaborating researchers or other research organizations. Anyone accessing these data from our server and using them in publications and/or other research products is asked to provide:

1. Full acknowledgement and citation of the contributing researcher(s) using the citation identified in the accompanying meta-data description.
2. Acknowledgment of access through the Muskegon Partnership program. We will try to develop a standardized boiler-plate for this.

Type 2: Project-Shared: These are data sets that individual collaborators are making available to other Muskegon collaborators, but do not want placed in the public domain. The data would usually be accessible via the controlled-access FTP site. All data collected as a part of the Muskegon Assessment and Modeling projects funded by GLFT should be made available in this format within 2 years of production (processing). Exceptions must be cleared with project leads (Stevenson at MSU [Michigan State University] or Wiley at UM [University of Michigan]). Collaborators using these data in publications and/or other research products agree to provide:

1. Full acknowledgement and citation of the contributing researcher(s) using the citation rules specified by the researcher and identified in the catalogue meta-data.
2. Acknowledgment of access through the Muskegon Partnership Program.
3. Agreement not to distribute the data in question to third parties without the written consent of the originating researcher.

Type 3: Team-Shared: These are data sets that individual researchers are interested in sharing with other collaborators in the context of a specific collaborative analysis. The data might be made accessible via the controlled-access FTP site, directly from the collaborating researcher. Most collected data should be released with a Type 3 designation within 6 months of processing. MRI collaborators using these data in publications and/or other research products are asked to provide:

1. Co-authorship and/or full acknowledgement and citation, details to be negotiated with the contributing researcher(s).
2. Acknowledgment of access through the Muskegon Partnership Program.
3. Agreement not to distribute the data in question to third parties without the written consent of the originating researcher.

Ordinarily data generated in these projects will progress from Type 3 to Type 1 availability on the following schedule:

1. Within 2 years of processing all data sets be will considered to be Type 2, exceptions require approval by project lead scientist.
2. By the end of the project period (Dec. 2007) all core data will be considered to be Type 1 (Public).

20 Integrated Modeling of the Muskegon River

Tools for Ecological Risk Assessment in a Great Lakes Watershed

Michael J. Wiley, Bryan C. Pijanowski,
R. Jan Stevenson , Paul Seelbach,
Paul Richards, Catherine M. Riseng,
David W. Hyndman, and John K. Koches

20.1 INTRODUCTION

The rapid pace and pervasiveness of landscape modification has made predicting watershed vulnerability to landscape change a key challenge for the twenty-first century. River ecosystems are, in particular, directly dependent on landscape structure and composition for their characteristic water and material budgets. Although it is widely acknowledged that landscape change poses serious risks to river ecosystems, quantification of past effects and future risks is problematic. Important issues of scale, hierarchy, and public investment intervene to complicate both assessment of current condition and the prediction of riverine responses to changes in landscape structure. In this paper we demonstrate how neural-net approaches to landscape change prediction can be coupled with river valley segment classification to provide a framework for integrated modeling and risk assessment across large-scale river ecosystems. Specifically we report on progress and techniques being employed in a collaborative risk assessment for the Muskegon River watershed, a large and valuable tributary of Lake Michigan.

Both watershed-based modeling and river classification have been proposed as methods of simplifying analysis in order to more efficiently protect river ecosystems (Hawkes 1975, Hudson et al. 1992, Maxwell et al. 1995, Wiley et al. 1997). Linking typical status and risk assessment models (e.g., bio-assessment protocols or predictive models, see Wiley et al. 2002) to explicit classification systems (Seelbach and Wiley 2005), however, remains a key methodological challenge. Ideally, a solution would provide both a spatially explicit classification system that simplifies the natural complexity of our rivers, and a method for coordinating suites of physical and biological models capable of predicting ecological status across a region and over time.

As a part of a large collaborative study (Stevenson et al. this volume) of the 2,600-square-mile Muskegon River watershed, we have recently developed a GIS-based approach using ecologically defined valley segment units (Seelbach et al. 1997, Seelbach and Wiley 2005, Seelbach et al. 2006) to integrate a state-of-the-art neural-net model (Landscape Transformation Model: LTM, Pijanowski 2000, 2002) with a variety of hydrologic and other models for the purpose of conducting rigorous integrated risk assessments at a watershed scale. The result is a modeling system, the Muskegon River Ecological Modeling System (MREMS), in which a variety of models can be used together to estimate risks to key watershed resources arising from various landscape change scenarios. Valley segment–scale ecological classification units (VSEC units; Seelbach and Wiley 2005) are used as an efficient and ecologically meaningful physical framework for organizing data exchanges among interacting models and stratifying model predictions. Output is remapped onto classification units to summarize and visually integrate spatially explicit forecasts of ecological status and future risk.

In this paper we provide a basic description of the structure of the MREMS system and detail the model linkage strategy we are employing. In addition, we provide preliminary examples of integrated assessment modeling based on the coupled execution of a series of land use change, hydrologic, loading, and biological response models from our Muskegon River studies.

20.1.1 METHODOLOGY

MREMS is a distributed modeling environment in which we are linking many different kinds of models to build a comprehensive picture of how the Muskegon River ecosystem functions (Figure 20.1). In many cases we are using several models of the same general phenomenon because often they employ different approaches, scales, or generate different types of useful output. Philosophically our approach is to recognize the inherent inaccuracies associated with all modeling and to favor redundancy by including many types of models, and modeling at multiple spatial scales.

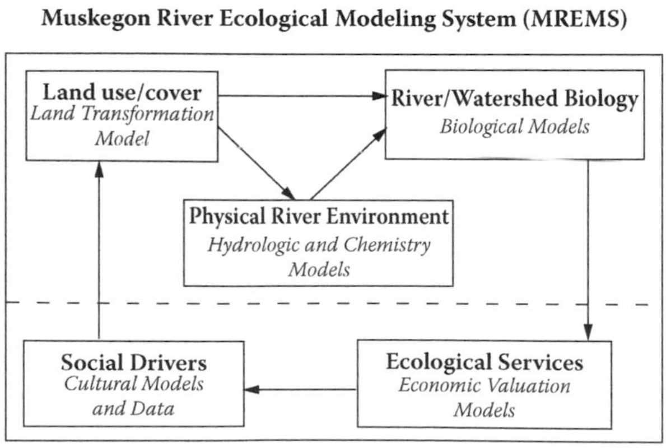

Muskegon River Ecological Modeling System (MREMS)

FIGURE 20.1   Schematic representation of the structure of MREMS components and typical execution order.

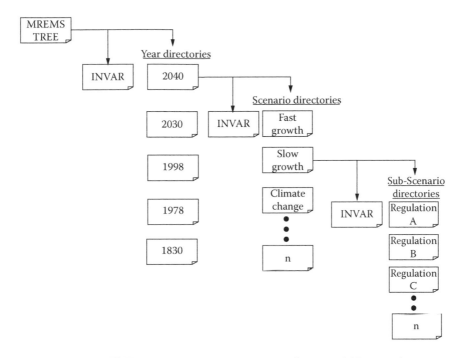

FIGURE 20.2   MREMS directory structure used to coordinate model input and output.

Therefore, MREMS can be best visualized as consisting of a suite of interacting sets of models, each focused on a particular aspect of the Muskegon River watershed environment. Integration occurs implicitly by requiring all models to either produce valley segment–scaled output or output at a higher hierarchical scale that can be used to drive finer-scaled models. Models that operate at reach or finer scales are required to aggregate output to produce generalizations for the valley segment in which they occur. MREMS system scaling adapts the classic hierarchy proposed by Frissell et al. (1986) and recognizes the following potential scales for model execution: basin, sub-basin, valley segment, reach (sub-vsec unit), channel habitat unit, cross-section.

Apart from its component models (see below) MREMS is essentially an explicit protocol and directory structure (Figure 20.2) that facilitates the linked execution of component models in a spatially explicit manner. MRI-VSEC version 1.1, a GIS (geographic information system) product, provides the spatial framework for referencing all input, output, and display of the component models in MREMS. Models communicate by placing appropriate identifiable output (*.txt or *.dbf) into a structured directory system that is organized into specific time frame (land cover sample year), problem context (scenario), and management option (sub-scenario) levels. At every level an INVAR (invariant) directory holds datasets, which are true for that and all lower levels of the directory space, as well as a subdirectory index, log, and other ancillary files (Figure 20.2). An MREMS run for a specific scenario involves the serial execution of a set of component models for each time frame, using scenario-specific, and sub-scenario-specific inputs and outputs. In many cases the output written by one model may be used as input by the next. Execution order is

TABLE 20.1

Component models linked in MREMS.

Model	Predicts	Type
MODFLOW	Groundwater flow	Sim
MRI_DARY	Groundwater inputs	GIS
HEC-HMS	Surface water flows	Sim
MRI_FDUR	Surface water flow frequencies	Linear
HEC-RAS	Surface water hydraulics	Sim
GWLF	Surface dissolved loads	Sim
MRI_LOADS	Surface dissolved loads	Regress
MRI_JTEMP	July water temp	Regress
Assessment Models		
All taxa	Fish/insect diversity	Regress
Sensitive taxa	Fish/insect diversity	Regress
EPT index	EPT taxa	Regress
Algal index	Algal status	Regress
Bioenergetic IBM		
Steelhead	Growth rate and survivorship	Sim
Salmon		Sim
Walleye		Sim
Biomass Composition		
Sport fishes	Kg/ha total mass	SEM, Regress
Total fishes	Kg/ha total mass	SEM
Sensitive fishes	Kg/ha total mass	SEM
Total algae	g/m^2	SEM
Filter-feeders	g/m^2	SEM
Grazing inverts	g/m^2	SEM

determined by data dependency. Typically, execution order starts with the generation of a land cover map (produced by LTM), followed by hydrologic, chemical loading, and ecological and biological models in that order (Figure 20.1).

20.1.2 MREMS COMPONENT MODELS

We have developed MREMS as an open system in which any type of model can in theory be used. At the present time we are working with suites of hydrologic, loading, and biological models (Table 20.2). These models represent much of the range in types of models used in natural resource planning contexts around the world. Some are simple GIS models, some linear statistical models that produce point estimates, and some are complex covariance structure models that describe both physical and biological processes. Several are large-scale dynamic simulation models (e.g., Hec-HMS, MODFLOW, several fisheries bioenergetic growth models). Beyond the neural net LTM, the most complex component models are the hydrologic simulations implemented using HEC-HMS, GWLF, and MODFLOW. A basinwide 15-minute time-step version of the HEC-HMS is now being refined. In MREMS it

TABLE 20.2

Example of future risk analysis by MREMS run for a fast-growth scenario. Multiple ecological responses predicted for 1998 to 2040 time frame comparison.

Site	% DQ[a]	Channel[b] Response	% SL[c]	% TDS[d]	Fish spp. loss
Cedar Creek	−13%	aggrade	+26%	+32%	3–4
Brooks Creek	−22%	aggrade	+72%	+20%	1–2
Main River @ Evart	0%	No change	+1%	+20%	2–3
Main River @ Reedsburg	0%	No change	+6%	+3%	0–1

[a] %DQ: percent change in dominant discharge (determines the size of the equilibrium channel).

[b] Channel response: expected response based on %DQ.

[c] %SL: percent increase in average daily sediment load (tons/day).

[d] %TDS: percent change in median total dissolved solids concentration (ppm).

uses a two-layer custom groundwater recharge routine to generate baseflow components, which are then added to and routed through the HEC-HMS surface water network. A scenario execution (see below) results in 20-year hydrographs being estimated for each of 56 model elements. These in turn are used to interpolate 20-year hydrographs for each of the 138 VSEC units in the Muskegon. HEC-HMS uses the SCS unit hydrograph approach to interpret LTM-projected land cover changes and produce resulting hydrographic predictions for the river system. The hydrographic projections are then used to drive a variety of other component models in MREMS.

The most critical model for running risk assessment scenarios in MREMS is the Land Transformation Model (Pijanowski et al. 2002), which provides us with changing land use distributions upon which many other component models react. LTM version 3 is a data-intensive neural net model that predicts land use change at the level of 30-m pixels across the landscape. Neural-net "imagined" landscapes, coupled with a standard 20-year climate scenario (1970–1990 observed temperatures and precipitation), and best available DEM and geology covers provide the physical template from which input parameters for constituent models are prepared. The Muskegon River drainage net itself (in the form of the VSEC framework) is then used to identify appropriate spatial strata for model parameterization and execution.

20.1.3 THE MRI-VSEC FRAMEWORK

For our model of the Muskegon watershed we have adapted the Michigan Rivers Inventory VSEC version 1.1 system (Seelbach et al. 1999, Seelbach and Wiley 2005) by correcting some minor mapping errors and transferring it to a 1:24000 scale channel cover based on 1978 (MDNR, MIRIS) air photos. We define ecological valley segments (VSEC units) as (variably) large sections of river channel that contain distinct, relatively homogeneous habitat conditions and biological assemblages. Higgins et al. (1999) referred to units of this type and scale as *fish macrohabitats*.

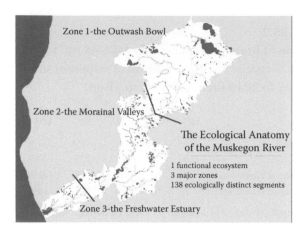

FIGURE 20.3 Watershed VSEC map providing the spatial framework for MREMS model linkage. Based on VSEC version 1.1 (Seelbach et al. 1997), all model output requires explicit referencing to one or more units.

Ecological valley segments combine elements of local valley and channel geomorphology with catchment hydrology, the two dominant forces shaping riverine habitat. In general, this approach is conceptually similar to the hydrogeomorphic (HGM) concept used in wetland assessment (Hauer and Smith 1998). The system identifies 138 distinct (contiguous) channel units in the Muskegon River ranging from first- to fifth-order channel segments (Figure 20.3). Major reservoirs and Muskegon, Houghton, Cadillac, and Higgins lakes are included as separate VSEC units. In MREMS, all models are required to provide model output referenced to one or more of the 138 segments. The resolution of the input and the scale at which the model executes (e.g., a single site, multiple sites in the segment, the entire watershed) is left to the individual model and modeler. Basic parameters for many landscape features (e.g., watershed land cover, surficial geology, elevation, basin size) are provided by the MREMS system for upper, mid-point, and lower nodes of each VSEC unit.

To illustrate the general MREMS methodology, Figure 20.4 shows data paths through MREMS used in a relatively simple coupling of 3 models (LTM, MRI_FDUR, and MRI_LOADS) used in proof-of-concept tests in 2002. LTM is the neural-net based landscape transformation model. MRI_FDUR, a hydraulic geometry-based model from the Michigan Rivers Inventory Program (Seelbach and Wiley 1997), predicts long-term flow duration curves for sites given landscape and climatic inputs. MRI_LOADS is an empirical nutrient-loading model that predicts instantaneous nutrient loads given land cover, geology, and catchment water yield. Sample sites are used to represent the entire VSEC unit in which they occur, based on the mapping criteria of ecological homogeneity, (Panel A, Figure 20.4). The VSEC unit ID number is used to geo-reference and query associated catchment, riparian buffer, and site scale databases to generate input parameters for component models (Panel B, Figure 20.4). Once output is generated by the MREMS component models, they are linked back to the VSEC unit ID and onto the VSEC spatial framework to produce channel maps with explicit model predictions for each of 138 VSEC channel segments. Panel C of Figure 20.4 shows the Muskegon VSEC unit map with

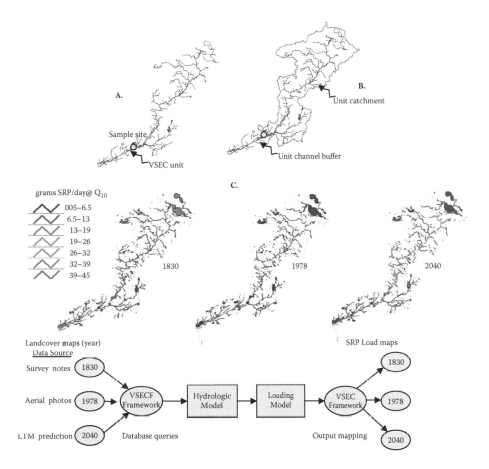

FIGURE 20.4 Illustration of model linkage in a simplified MREMS run. (**See color insert after p. 162.**) See text for detailed explanation. Panel A illustrates the representation of a sample site location by the mapped VSEC unit in which it occurs. Database information for the unit's upstream catchment, local riparian buffer, and other attributes are linked to the site via the VSEC unit ID (Panel B). Panel C illustrates information flow and final model output mapping on the VSEC units for a simple run linking land cover data, MRI-FDUR (a hydrologic model), and MRI-LOADS (an empirical nutrient-loading model) to predict daily phosphate loads at flood flows (Q_{10} = 10% annual exceedence discharge for the VSEC unit).

predicted phosphate loading over time. The illustrated 2040 scenario gives expected loads at the 10% annual exceedence discharge if high rates of urbanization observed in the 1990s were to continue to the year 2040.

20.2 PRELIMINARY RESULTS FOR A RAPID DEVELOPMENT SCENARIO

Full implementation and parameterization of the MREMS modeling system is not scheduled to be complete until late 2007, awaiting the completion of field studies across the Muskegon basin. Nevertheless a number of preliminary runs have already

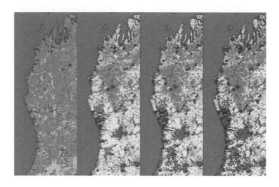

FIGURE 20.5 Sequence of land cover scenarios used to drive preliminary MREMS executions. **(See color insert after p. 162.)** The source for 1820 is MDNR digitized GLO notes; 1998 through 2040 are LTM neural-net projections from base 1978 MIRIS air photo coverage.

been made, both to calibrate and evaluate component models and to refine linkage protocols. These early runs use LTM projections assuming a 1990s rate of growth and therefore provide a kind of "worst likely case" development scenario for the basin. These runs are already proving useful in focusing current conservation and restoration activities. The spatially explicit nature of the MREMS system identifies those segments of the rivers that are most at risk from rapid development and likely patterns of land use change.

Regional LTM projections for the year 2040 using a fast growth scenario suggest that most of the additional urbanization in the basin will occur along the Lake Michigan–U.S. 131 corridor, and secondarily along other major transportation corridors across the Muskegon watershed (Figure 20.5). LTM-coupled HEC-HMS and GWLF runs provide a basis for examining both direct hydrologic responses and indirect hydrologic effects by driving other model impacts on water quality, sediment transport, potential channel geometry, and ultimately the response of biological communities. For example, HEC-HMS output for Cedar Creek (a key lower river tributary) showed a small but important hydrologic response to the 1998 versus 2040 landscape configuration using identical precipitation forcing. Even though Cedar Creek is predominantly driven by groundwater inputs, the MREMS run suggests anticipated increases in impervious surface will increase event peak discharge rates in the channel by nearly 100%, but baseflow response will be minimal. Using the modeled hydrographic data in dominant discharge analyses in turn indicates that sediment transport in Cedar Creek is likely to increase by 32% on an annual basis. Further, resulting changes in the transport regime are likely to lead to channel aggradations and loss of important fish habitat (Table 20.2). Coupled biological models suggest extirpation of 2 to 3 of the 10 or so species currently found in this tributary. Similar but somewhat more dramatic impacts were predicted for Brooks Creek, an adjacent and more agriculturally developed watershed. In Brooks Creek, larger impacts on hydrology and sediment loading were predicted, but biological models predicted fewer species would be lost compared to adjacent Cedar Creek. This difference in magnitude of the expected biological response reflects differences in the

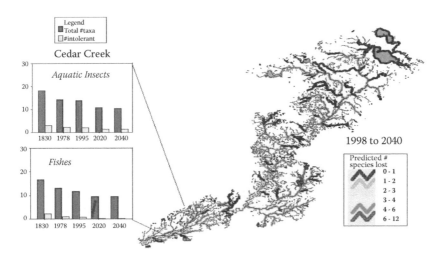

FIGURE 20.6   Changes in biological diversity predicted in response to land cover change predicted in the "fast growth" LTM scenario. **(See color insert after p. 162.)**

importance of groundwater loading in the basins, leading to subsequent differences in temperature and the fish community structure. Nutrient-loading models likewise indicate large increases in nitrogen and phosphorus export from these tributaries (see Figure 20.4).

Regression models predicting biological community response (see Wiley et al. 2003) required, as input parameters, estimates of TDS (total dissolved solids) concentration, baseflow yield, catchment area, and the percentage of the catchment in urban and agricultural land cover. Adjusting inputs based on LTM, hydrologic, and loading model predictions, total diversity and number of intolerant species were predicted for each VSEC unit in the river system. Mapping the change in diversity across the basin provides a spatially explicit map of the risk of species loss due to predicted landscape development (Figure 20.6). Since combining historical data, air photo–based GIS coverages, and LTM predictions yields a series of land cover maps, MREMS can be used to produce a sequence of hindcasts and forecasts that model the trajectory of biodiversity in any VSEC unit of interest.

For example, in our early MREMS runs the fast-development scenario described above affects biological diversity principally in the main stem and lower river tributaries. Most of the main stem downstream of Evart is predicted to lose 1 to 2 species. The segment immediately below Cedar Creek (N. Branch lower Muskegon) and Cedar Creek itself were the most seriously threatened. Cedar Creek is predicted to lose 3 to 4 species and the N. Branch Muskegon (in the Fish and Game Area) 4 to 6 species. These declines are relative to modeled diversity, using the 1998 land cover configuration. As can be seen from the insets in Figure 20.6, this decline is a part of a trend in declining diversity, which the MREMS analysis suggests began with the onset of heavy settlement in the nineteenth century. Both aquatic insects and fish diversity decline over time with intolerant fishes in particular being vulnerable.

20.2.1 Discussion

Although final implementation and risk assessment modeling with MREMS lies ahead, limited runs to date are already proving useful in both watershed restoration planning and study design contexts. The spatially explicit nature of the modeling system facilitates visualization and communication about potential risks to this important river resource. In particular, Cedar Creek in Muskegon County has repeatedly emerged as a tributary system clearly at risk from development. These results have already led to increased attention and conservation planning efforts for Cedar Creek. These include a fisheries habitat inventory being directed by the NRCS (Natural Resources Conservation Service), a volunteer–university collaboration to develop sediment-loading functions for Cedar Creek, a new MDNR-MDEQ (Michigan Department of Natural Resources and Michigan Department of Environmental Quality) collaborative modeling aimed at identifying potential hydrologic storage and baseflow protection BMPs, and as a part of our MREMS calibration work, we have increased the density of automated gauging installations in an effort to improve the precision of our hydrologic predictions.

Our early experiences with Cedar Creek arose out of early proof-of-concept modeling runs completed in 2003. Ultimately, when we run the final basinwide risk assessments for which MREMS is designed we will be evaluating various management scenarios developed by a focus group of collaborating Muskegon watershed stakeholders. At a stakeholders' workshop in August 2002 they identified three major types of scenarios that they would like to evaluate using the MREMS system. These categories include land management scenarios (e.g., evaluating different sized riparian setbacks, evaluating effects on alternate rates and sites of development), hydrologic management scenarios (e.g., evaluating dam and lake level control effects, examining the effect of wetland losses and protection on river hydrology), and sediment/erosion management scenarios (e.g., where is bank erosion and aggradation being affected by development? Where is bank stabilization a useful strategy?). A full list of the MREMS risk assessment scenarios developed at the stakeholder workshop are available at http://mwrp.net/mrems/.

20.2.2 Future Plans and Benchmarks

Modern GIS systems provide the appropriate technology for blending bottom-up site-based modeling (and sampling) with top-down regionalization and mapping approaches (see review by Seelbach et al. 2001). We are demonstrating that advanced landscape-transformation models can be systematically linked to a landscape-cognizant, ecologically interpreted river segment classification system, to provide an effective spatial framework for both sampling inventory and spatially explicit modeling of river status and risk with respect to landscape alterations. The value of this approach lies principally in (1) the orchestration of integrated model-based assessments by standardizing units of data exchange instead of scales of parameterization and analysis, and (2) the resulting spatially explicit visualization of the complex products of landscape and other environmental change. Beginning and ending with maps, while maintaining the rigor of process-based and site-specific modeling, our approach brings the capability of detailed technical information processing to the public in a fashion that is relative easy to comprehend. Its value can, in part, be measured by the almost immediate utility of early output in identifying threatened subcatchments like Cedar Creek in the Muskegon watershed.

ACKNOWLEDGMENTS

The Great Lakes Fisheries Trust is funding work on the MREMS modeling for the Muskegon River. The Wege Foundation has supported much of the land cover data analysis used in the LTM neural-net model. Parts of the work referred to here were also funded by grants to the various coauthors from the U.S. Environmental Protection Agency, the National Science Foundation, and the Michigan Department of Natural Resources. Scores of students and collaborators from across the Great Lakes region have contributed and continue to participate in the Muskegon Watershed Research Partnership (www.mwrp.net) and to all we are sincerely grateful.

REFERENCES

Frissell, C. A., W. J. Liss, C. E. Warren, and M. D. Hurley. 1986. A hierarchical framework for stream habitat classification: Viewing streams in a watershed context. *Environmental Management* 10:199–214.

Hauer, F. R., and R. D. Smith. 1998. The hydrogeomorphic approach to functional assessment of riparian wetland: Evaluating impacts and mitigation on river floodplains in the U.S.A. *Freshwater Biology* 40:517–530.

Hawkes, H. A. 1975. River zonation and classification. In *River Ecology,* ed. B. A. Whitton. Berkeley, CA: University of California Press, 312–374.

Hawkins, C. P., R. H. Norris, J. Gerritsen, R. M. Hughes, S. K. Jackson, R. K. Johnson, and R. J. Stevenson. 2000. Evaluation of the use of landscape classifications for the prediction of freshwater biota: Synthesis and recommendations. *Journal of the North American Benthological Society* 19:541–556.

Higgins, J., M. Lammert., and M. Bryer. 1999. *Including aquatic targets in ecoregional portfolios: Guidance for ecoregional planning teams. Designing a geography of hope.* Update #6. Arlington, VA: The Nature Conservancy.

Hudson, P. L., R. W. Griffiths, and T. J. Wheaton. 1992. Review of habitat classification schemes appropriate to streams, rivers, and connecting channels in the Great Lakes drainage basin. In *The Development of an aquatic habitat classification system for lakes,* ed. W.-D. N. Busch and P. G. Sly. Boca Raton, FL: CRC Press, 73–107.

Maxwell, J. R., C. J. Edwards, M. E. Jensen, S. J. Paustian, H. Parrott, and D. M. Hill. 1995. *A hierarchical framework of aquatic ecological units in North America (Nearctic Zone).* St. Paul, MN: U.S. Department of Agriculture Forest Service, North-Central Forest Experiment Station, General Technical Report NC-17.

Pijanowski, B. C., S. H. Gage, and D. T. Long. 2000. A land transformation model: Integrating policy, socioeconomics and environmental drivers using a geographic information system. In *Landscape ecology: A top-down approach,* ed. Larry Harris and James Sanderson. Boca Raton, FL: Lewis Publishers.

Pijanowski, B. C., D. G. Brown, G. Manik, and B. Shellito. 2002. Using artificial neural networks and GIS to forecast land use changes: A land transformation model. *Computers, Environment and Urban Systems.* 26(6):553–575.

Seelbach, P. W., M. J. Wiley, J. C. Kotanchik, and M. E. Baker. 1997. *A landscape-based ecological classification system for river valley segments in lower Michigan (MI-VSEC version 1.0).* Michigan Department of Natural Resources, Fisheries Research Report 2036. Ann Arbor: Michigan Department of Natural Resources.

Seelbach, P. W., and M. J. Wiley. 1997. *Overview of the Michigan Rivers Inventory Project*. Michigan Department of Natural Resources, Fisheries Technical Report 97-3, Ann Arbor: Michigan Department of Natural Resources.

Seelbach, P. W., and M. J. Wiley. 2005. An initial landscape-based system for ecological assessment of Lake Michigan tributaries. In *State of Lake Michigan: Ecology, health and management*, ed. T. Edsall and M. Munawar. Aquatic Ecosystem Health and Management Society, 559–581.

Seelbach, P. W., M. J. Wiley, P. Soranno, and M. Bremigan. 2001. Aquatic conservation planning: Predicting ecological reference ranges for specific waters across a region from landscape maps. In *Concepts and applications of landscape ecology in biological conservation*, ed. K. Gutzwiller. New York: Springer-Verlag.

Stevenson, R. Jan, Michael J. Wiley, Stuart H. Gage, Vanessa L. Lougheed, Catherine M. Riseng, Pearl Bonnell, Thomas M. Burton, R. Anton Hough, David W. Hyndman, John K. Koches, David T. Long, Bryan C. Pijanowski, Jiaquo Qi, Alan D. Steinman, and Donald G. Uzarski. 2007. Watershed science: Essential, complex, multidisciplinary and collaborative. Boca Raton, FL: CRC Press, 231–245.

Wiley. M. J., S. L. Kohler, and P. W. Seelbach. 1997. Reconciling landscape and site based views of stream communities. *Freshwater Biology* 37:133–148.

Wiley, M. J., P. W. Seelbach, and K. E. Wehrly. 2003. Regional ecological normalization using linear models: A meta-method for scaling stream assessment indicators. In *Biological response signatures: Patterns in biological integrity for assessment of freshwater aquatic assemblages*, ed. T. Simon. Boca Raton, FL: CRC Press, 201–223.

21 Watershed Management Practices for Nonpoint Source Pollution Control

Shaw L. Yu, Xiaoyue Zhen, and Richard L. Stanford

21.1 INTRODUCTION

Water quality protection is very important to maintaining human health and ecological integrity. A sustainable use of water resources is especially important in China due to the rapid economic growth and the accompanying urbanization in recent years. Traditional control technology tends to emphasize the *collection and treatment approach*. In recent years, *control at the source* is widely recognized as a more cost-effective alternative. Because source control techniques impact on all sectors of a society, socio-economic factors become important in the implementation of control measures.

The *watershed protection approach* (WPA) is a strategy for protecting and restoring aquatic ecosystems and protecting human health. This strategy is based on the notion that many water quality and ecosystem problems are best solved at the watershed level rather than at the individual water body or discharger level. WPA is an effective way to protect water quality while at the same time promoting a partnership approach forged by all stakeholders so that a balanced scheme can be realized, which will on the one hand protect the water resource in the watershed, and on the other hand allow reasonable development in the watershed.

21.1.2 EFFECTS OF URBANIZATION ON THE WATER ENVIRONMENT

The environmental effects of urbanization are well known. However, most of the attention given to the environmental effects of urbanization deal with air pollution from the increased number of automobiles, water pollution from the increased density of population, and solid wastes. Only now is there increasing attention being paid to the effects of urbanization on natural resources. We have tended to look at the problems associated with such things as water supply only from the *demand side* related to increased population and not from the *supply side*, considering the effect that urbanization has on diminishing the supply.

The major impact of urbanization on the water environment can be summarized as follows:

- Hydrology—higher flood peaks, larger runoff volume, faster flood flows, less evapotranspiration, and less groundwater recharge.

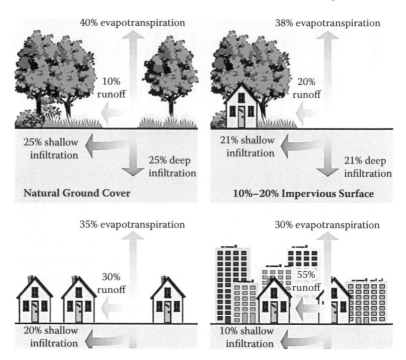

FIGURE 21.1 Effect of urbanization on hydrology. (Federal Interagency Stream Restoration Working Group (FISRWG). 1998. Stream corridor restoration: Principles, processes, and practices. GPO Item No. 0120-A; SuDocs No. A 57.6/2:EN 3/PT.653. ISBN-0-934213-59-3. http://www.usda.gov/technical/stream_restoration/.)

- Water quality—larger wastewater volumes, enhanced sediment and erosion processes, and stormwater runoff pollution.
- Aquatic biological integrity—habitat loss, biodiversity, toxicity, and so forth.

21.1.2.1 Hydrology

The porous and varied terrain of natural landscapes like forests, wetlands, and grasslands trap rainwater and snowmelt and allow it to slowly filter into the ground. Infiltrating water replenishes aquifers, and runoff tends to reach receiving waters gradually. In contrast, nonporous and uniformly sloping urban landscapes, which include features like roads, bridges, parking lots, and buildings, prevent runoff from slowly percolating into the ground. Figure 21.1 shows the relationship between various degrees of urbanization and the hydrologic cycle. It is clear that the predominant effect is to reduce the amount of infiltration and route the water into runoff.

Urban developers install storm sewer systems that quickly channel runoff from impervious surfaces. When this collected runoff is discharged into streams, large volumes of quickly flowing runoff erode the banks, damage streamside vegetation, and widen stream channels. In turn, this process results in lower water depths during non-storm periods and higher than normal water levels during wet weather periods (i.e., *flashiness*), increases in sediment loads, and higher water temperatures, and so forth.

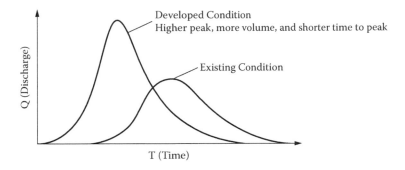

FIGURE 21.2   Hydrological impact of urbanization.

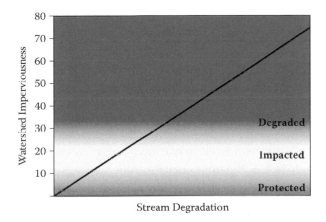

FIGURE 21.3   Relationship between imperviousness in a watershed and stream quality. (**See color insert after p. 162.**) (T. Scheuler. 1994 "The Importance of Imperviousness," Center for Watershed Protection, Columbia, MD., *Watershed Protection Techniques* 1(3):101.)

Figure 21.2 illustrates an example of the effect of urbanization on the rainfall-runoff process, that is, higher flood peaks and shorter time of travel for the stormwater runoff.

21.1.2.2 Water Quality and Ecological Impacts

In addition to adverse effects on hydrology, it is well established that urbanization has significant adverse effects on the quality of both surface and groundwater.

Aquatic life cannot survive in urban streams severely affected by urban runoff. Figure 21.3 shows the relationship between the percentage of imperviousness in a watershed and the degree of stream degradation that can be expected. It is clear that once a watershed reaches roughly 30% impervious surfaces, significant degradation of streams in terms of water quality and ecological health in that watershed can be expected.

The relationship between imperviousness in a watershed and stream quality is based on empirical studies in the United States. Table 21.1 shows the results of a nationwide study of the quality of stormwater in the United States. These results were

TABLE 21.1

Median stormwater pollutant concentrations for all sites by land use.

Pollutant	Residential Median	Residential COV	Mixed Land Use Median	Mixed Land Use COV	Commercial Median	Commercial COV	Open/ Nonurban Median	Open/ Nonurban COV
BOD$_5$ (mg/L)	10.0	0.41	7.8	0.52	9.3	0.31	—	—
COD (mg/L)	73	0.55	65	0.58	57	0.39	40	0.78
Total soluble solids (mg/L)	101	0.96	67	1.14	69	0.85	70	2.92
TKN (µg/L)	1900	0.73	1288	0.50	1179	0.43	965	1.00
NO$_2$-N+NO$_3$-N (µg/L)	736	0.83	558	0.67	572	0.48	543	0.91
Total Phosphorus (µg/L)	383	0.69	263	0.75	201	0.67	121	1.66
Sol. Phosphorus (µg/L)	143	0.46	56	0.75	80	0.71	26	2.11
Total Pb (µg/L)	144	0.75	114	1.35	104	0.68	30	1.52
Total Cu (µg/L)	33	0.99	27	1.32	29	0.81	—	—
Total Zn (µg/L)	135	0.84	154	0.78	226	1.07	195	0.66

Note: COV: coefficient of variation = standard deviation/mean; BOD = biological oxygen demand; COD = chemical oxygen demand; TKN = total Kjeldahl nitrogen.

Source: U.S. Environmental Protection Agency (USEPA). 1983. *Final Report, Nationwide Urban Runoff Program.* Washington, DC: U.S. Environmental Protection Agency.

obtained through the U.S. Environmental Protection Agency–sponsored Nationwide Urban Runoff Program (U.S. Environmental Protection Agency [USEPA] 1983).

In Table 21.1 it can clearly be seen that there was a significantly greater concentration of pollutants in the stormwater from residential land use, mixed land use, and commercial land use than from open/nonurban land use. This greater concentration, combined with the increased discharge to streams in urban areas, results in greatly increased loadings of pollutants in streams and other receiving waters. For example, highway construction impacts include excessive sediment yield during construction and runoff pollution from pavements and right-of-ways. For example, hydrologic changes due to site cleaning, grading, increased imperviousness, and landscape maintenance can cause stream channel instability, which could lead to stream bank erosion and habitat degradation (Federal Highway Administration [FHWA] 2000).

21.2 WATERSHED MANAGEMENT STRATEGY AND PRACTICES

21.2.1 THE TMDL CONTROL STRATEGY

The TMDL (total maximum daily load) of a water body is defined as the total allowable loading of a pollutant from all sources, point and nonpoint, entering the water body so that the water quality standards are not violated. For a water body the TMDL can be expressed as:

$$TMDL = LC = WLA + LA + MOS \tag{21.1}$$

where TMDL = total maximum daily load; LC = loading capacity of the water body; WLA = portion of the TMDL allocated to point sources; LA = portion of the TMDL allocated to nonpoint sources; and MOS = margin of safety or uncertainty factor.

The necessary components of a TMDL process should include the following:

- Selection of the pollutant or pollutants to consider
- Estimation of the water body assimilative capacity
- Estimation of the pollution from all sources, including background
- Simulation of the fate and transfer of pollutants in the water body and the determination of total allowable load under critical or design conditions
- Allocation of the allowable load among all sources in a manner enabling water quality standards to be achieved
- Consideration of seasonal variations and uncertainties
- Inclusion of public and stakeholder participation

The TMDL process is currently the main driving force sustaining the water quality control efforts throughout the United States. For example, in Virginia there were more than 80 TMDL studies scheduled during the past decade. One of these studies was conducted by the University of Virginia (Yu and Zhang 2001). The study involved the development of a control strategy for nitrate pollution for the Muddy Creek watershed in northwestern Virginia. The nitrate TMDL was first determined based on the *assimilative capacity* of Muddy Creek with respect to nitrate. The total permissible loads were then distributed among various point and nonpoint sources in the watershed. Different load reduction scenarios were generated and compared. A final load allocation scheme was selected after much discussion among the stakeholders involved in the TMDL process.

21.2.2 BEST MANAGEMENT PRACTICE (BMP) TECHNOLOGY

Best management practices (BMPs) are structural or nonstructural practices designed for the removal or reduction of nonpoint source pollution. Examples of these practices include storage facilities such as detention ponds, infiltration facilities such as infiltration trenches and porous pavements; vegetative practices such as grassed filter strips and swales, and constructed wetlands. More recently, the low-impact development (LID) type of BMP has received a great deal of attention. These BMPs and those that are especially appropriate for application in urban areas and highway construction are briefly discussed in the following sections.

21.3 PRACTICES FOR ECO-FRIENDLY URBAN DEVELOPMENT AND HIGHWAY CONSTRUCTION

21.3.1 LOW-IMPACT DEVELOPMENT (LID) TECHNIQUES

Low-impact development (LID) techniques are simple and effective, and are significantly different from conventional engineering approaches, which emphasize the piping of water to low spots removed from the development area as quickly as possible. Instead, LID uses *micro-scale techniques* (sometimes known as *ultra-urban techniques*) to manage precipitation as close to where it hits the ground as possible. The basic principles of low-impact development include (Coffman 2001):

- Restore/conserve natural hydrologic processes
- Increase flow paths
- Hydraulically disconnect impervious surfaces
- Upland phytoremediation
- Disburse runoff
- Unique watershed storage
- Minimize imperviousness
- Multifunctional landscaping
- Integrated micro-scale management
  - Retain
  - Detain
  - Recharge
  - Treat

One of the primary goals of LID design is to reduce runoff volume by infiltrating rainfall water to groundwater, evaporating rainwater back to the atmosphere after a storm, and finding beneficial uses for water rather than exporting it as a waste product down storm sewers. The result is a landscape functionally equivalent to predevelopment hydrologic conditions, which means less surface runoff and less pollution damage to lakes, streams, and coastal waters. LID practices include such techniques as bioretention cells or rain gardens, grass swales and channels, vegetated rooftops, rain barrels, cisterns, vegetated filter strips, and permeable pavements. Many of these techniques both reduce runoff volume and filter pollutants from water before it is discharged into receiving watercourses. Several of the most commonly used LID practices are briefly described below.

21.3.2 Bioretention

One of the key LID techniques is bioretention (sometimes referred to as *rain gardens*). Bioretention is a terrestrial-based (upland as opposed to wetland), water quality and water quantity control practice using the chemical, biological, and physical properties of plants, microbes, and soils for removal of pollutants from stormwater runoff. Some of the processes that may take place in a bioretention facility include: sedimentation, adsorption, filtration, volatilization, ion exchange, decomposition, phytoremediation, bioremediation, and storage capacity (Prince George's County 2002). Figure 21.4 shows a typical bioretention system.

Bioretention systems are more than simply creative landscaping. They are engineered systems that have been designed and installed to promote the biological, physical, and chemical treatment of stormwater runoff, as well as to promote the infiltration of stormwater runoff in order to help restore the character of the natural hydrologic cycle of the area. Bioretention cells are comprised of six basic components (U.S. EPA 2000).

FIGURE 21.4 Typical rain garden bioretention system.

These are:

- Grass buffer strips that reduce runoff velocity and filter particulate matter.
- Sand bed that provides aeration and drainage of the planting soil and assists in the flushing of pollutants from soil materials.
- Ponding area that provides storage of excess runoff and facilitates the settling of particulates and evaporation of excess water.
- Organic layer that performs the function of decomposition of organic material by providing a medium for biological growth (such as microorganisms) to degrade petroleum-based pollutants. It also filters pollutants and prevents soil erosion.
- Planting soil that provides an area for stormwater storage and nutrient uptake by plants. Often the planting soils contain some clays, which adsorb pollutants such as hydrocarbons, heavy metals, and nutrients.
- Vegetation (plants) that function in the removal of water through evapotranspiration and pollutant removal through nutrient cycling.

Laboratory and some limited field tests have shown good removal capabilities of some pollutants, such as 80%–90% for total suspended solids (TSS); 40%–50% for total phosphorus (TP), and 50%–90% for heavy metals (Federal Highway Administration [FHWA] 2000, Yu and Wu 2001). One significant advantage of bioretention cells as water management measures in urban areas is the fact that they can be designed as part of the urban or highway landscape and are relatively low cost in terms of construction and maintenance. Figure 21.5 shows the nitrogen cycle that occurs in a typical bioretention cell.

21.3.3 GRASSED SWALES

Swales are grassy depressions in the ground designed to collect stormwater runoff from streets, driveways, rooftops, and parking lots. Two general types of grassed swales are generally designed: (1) a dry swale, which provides water quality benefits by facilitating stormwater infiltration, and (2) a wet swale, which uses residence

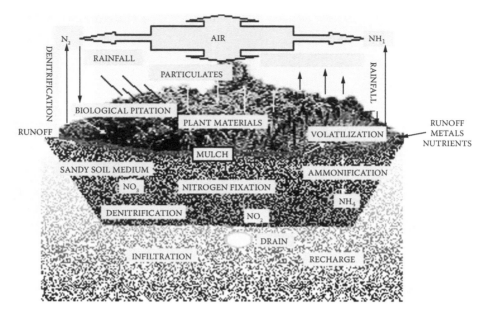

FIGURE 21.5   Bioretention nitrogen cycle. **(See color insert after p. 162.)**

time and natural growth to treat stormwater prior to discharge to a downstream surface water body. Both dry and wet swales demonstrate good pollutant removal, with dry swales providing significantly better performance for metals and nitrate (FHWA 2000). The primary pollutant removal mechanism is through sedimentation of suspended materials. Therefore, suspended solids and adsorbed metals are most effectively removed through a grassed swale. Dry swales typically remove 65% of total phosphorus (TP), 50% of total nitrogen (TN), and between 80% and 90% of metals (Yu and Kaighn 1995). Wet swale removal rates are closer to 20% of TP, 40% of TN, and between 40% and 70% of metals. The total suspended solids (TSS) removal for both swale types is typically between 80% and 90%. In addition, both swale designs should effectively remove petroleum hydrocarbons based on the performance reported for grass channels (FHWA 2000).

21.3.4 ECOLOGICAL DETENTION SYSTEMS

In general, LID technologies are applicable for small-scale contributing areas. For example, once the drainage area to a bioretention cell exceeds about 0.3 hectares, it may not be practical to use bioretention due to capacity limitations. In these cases, larger systems such as ponds and wetlands are generally used to treat stormwater (Center for Watershed Protection 1996). The larger stormwater management structures include retention basins, detention basins, extended-detention basins, and enhanced extended-detention basins.

An extended-detention basin is usually dry during non-rainfall periods. An enhanced or ecological extended-detention basin has a higher efficiency than an extended-detention basin because it incorporates a shallow marsh, or wetland system, in its bottom. The wetland provides additional pollutant removal through wetland

FIGURE 21.6   Stormwater treatment wetland. (Virginia Department of Conservation and Recreation (VADCR). 1999. *Virginia Stormwater Management Handbook*. http://www.dcr.state.va.us/sw/stormwat.htm.)

plant uptake, absorption, physical filtration, and decomposition. The wetland vegetation also helps to reduce the resuspension of settled pollutants by trapping them (Virginia Department of Conservation and Recreation [VADCR] 1999). Figure 21.6 shows a typical stormwater treatment wetland system, with a forebay and a marsh area. Wetland treatment systems differ from conventional detention and retention treatment systems by being shallow (generally less than 30 cm deep), having a large quantity of emergent and suspended aquatic vegetation, and emphasizing slow-moving, well-spread flow (Maestri et al. 1988).

Ecological processes inherent in such wetland stormwater treatment systems include sedimentation, adsorption of pollutants by sediments vegetation and detritus, physical filtration, microbial uptake of pollutants, uptake of pollutants by wetland plants, uptake of pollutants by algae, and other physical-chemical processes. The combination of ecological processes makes wetlands relatively effective in removing pollutants normally found in stormwater.

21.4 THE BIG CHALLENGE AHEAD

It is clear that China is increasing its rate of urbanization. A December 2002 news article (People's Daily 2002) indicated that the urbanization level in China stands at about 37%, roughly 10 percentage points lower than its industrialization level, and that China is able to upscale its urbanization level one to two percentage points every year, and finally reach a level of over 50 percent by the year of 2020. A later new article (People's Daily 2003) reported that China's urbanization level had risen from 10.6% in 1949 to 17.92% in 1978, and finally to 39.1% in 2002, and indicated that China will strive to harmonize economic growth, environmental protection, and urban development in the urbanization process, especially in the coming 20 years.

The Chinese Ministry of Water Resources (2003, p. 4) issued a report that stated, in part:

In some regions, overdraft of groundwater has caused serious regional declines in the groundwater table, creating a series of ecological problems such as large-scale land subsidence, reduction of ecological oasis, environment deterioration. Also, the problems of water pollution and soil and water loss are very serious in China. Flood disasters, water shortage, water pollution, and soil and water loss have seriously hampered the harmonious development of population, resources, environment and the economic society in China, and they have been the main constraints to the development of the Chinese economic society. Therefore, China must implement the sustainable water resources development strategy, strengthen the construction of water infrastructure, consolidate the building up and protection of the ecological environment, conserve and protect water resources, control water pollution, improve the ecological environment, promote the sustainable use of water resources, and safeguard the sustainable development of the economic society.

With China's rapid economic growth, the pressure for industrial and other urban development is very intense. Consequently, this is a critical time for China to develop a comprehensive plan for protecting its waters and ecosystems, while allowing carefully planned developments to move forward. The task is obviously very challenging, yet extremely important.

21.4.1 Implementation Issues in China

21.4.1.1 Regulatory Framework

In order to efficiently reach set goals for watershed water quality protection, a regulatory framework is needed. Requiring eco-friendly engineering practices for government-sponsored engineering projects, such as highway construction, is a very good strategy, but for privately sponsored construction projects, such as shopping malls and residential sites, the developers might not feel "obliged" to build and maintain BMPs. The regulatory framework could be established at either the central or local government level or both. Tax benefits could be used as a motivational tool.

21.4.1.2 Cost and Maintenance

One of the key issues in BMP implementation is: Who should pay for the construction and maintenance costs associated with the BMPs? In the United States, for public construction projects including road building, BMP cost is part of the overall construction cost and the responsible agency (e.g., transportation departments in the case of highway construction) would maintain the facilities. For private projects, the developer would construct the BMPs and the users (e.g., homeowners' associations in the case of residential developments) would be responsible for the maintenance costs.

BMP costs depend largely on the type of BMP and many other site-specific factors such as land value, labor and material costs, and so forth. The FHWA report in 2000 cited some preliminary costs for BMPs. For example, a bioretention cell system could cost about $25,000 per impervious hectare area served. On the other hand, swales and vegetative filter strips would cost much less, about $4,000 to $5,000 per impervious hectare served.

21.4.1.3 Technical Issues

Because nonpoint pollution problems are very site specific, there is virtually no one-size-fits-all type of approach in controlling NPS pollution. Rainfall, and therefore runoff characteristics in China are quite different from those in the United States. Factors such as topography, soil, agricultural practices (e.g., tea and fruit gardens are prevalent in some parts of China), climate, and so forth, all impact the selection and the design of BMPs that are appropriate for China. Some of the most important design-related issues are:

- What should be the design frequency for storms? (In the United States, a 10-year frequency is commonly used for runoff quantity control, whereas a 2-year or lesser storm frequency is used for quality control.)
- Should the control of the "first-flush" of the runoff (usually the first 0.5 in or 13 mm of runoff volume), which has been adopted in many states in the United States, be considered in China?
- Should the underground type of BMPs, such as bioretention cells, vault structures, and sand filters, be considered as preferred BMPs in China? These BMPs require little space and would be less vulnerable to vector problems, which should be a concern in warm-weather regions in China.
- How to deal with combined sewage and stormwater conveyance system regions. Current BMP design guides target stormwater runoff for treatment. For a mixture of stormwater and wastewater, the BMP must be designed accordingly. One possible design may be the BMP *treatment train*, which can include pretreatment processes capable of treating high-concentration pollutants.

21.4.1.4 Other Issues

Other important issues relating to a full-scale BMP implementation include: special provisions for certain sectors in the society (e.g., BMP implementation for the agricultural sector, especially farmers), partnership with environmental groups, public education strategies, and so forth.

21.5 CONCLUSIONS AND RECOMMENDATIONS

Urbanization, including highway construction, could cause significant negative impact on the aquatic ecosystem. There are a number of engineering practices, called BMPs, which can be employed to mitigate these negative impacts. BMPs such as bioretention cells, vegetative buffer strips and swales, and constructed wetlands can be integrated into the landscape and therefore provide both water quality management and aesthetic benefits. The full implementation of BMPs in watersheds requires a well-planned strategy, which needs to address issues such as a regulatory framework, cost and maintenance, and technical and other issues.

As China continues to urbanize, it is important to consider the environmental aspects, including the increased pollution and demands on natural resources. However, in addition to controlling gross pollution, China needs to also ensure that adverse effects on natural resources, especially water resources, are minimized.

New developments, including highway construction, should attempt to maintain the volume of runoff at predevelopment levels by using structural controls and pollution prevention strategies. Plans for the management of runoff, sediment, toxins, and nutrients can establish guidelines to help achieve both goals. Management plans that include low-impact development measures protect sensitive ecological areas, minimize land disturbances, and retain natural drainage and vegetation.

REFERENCES

Center for Watershed Protection. 1996. Dry weather flow in urban streams. Technical Note No. 59. *Watershed Protection Techniques* 2(1):284–287.

Chinese Ministry of Water Resources. 2003. China Country Report on Sustainable Development - Water Resources. February 24, 2003. http://www.chinawater.gov.cn/english1/pdf/china2003.pdf (accessed August 24, 2007).

Coffman, L. 2001. Low-impact development—a decentralized stormwater management approach to a functional ecosystem-based design. http://www.co.pg.md.us/Government/AgencyIndex/DER/PPD/LID/pdf/LID_Art2.pdf (accessed August 24, 2007).

Federal Highway Administration (FHWA). 2000. *Stormwater best management practices in an ultra-urban setting: Selection and monitoring.* http://www.fhwa.dot.gov/environment/ultraurb/index.htm (accessed August 24, 2007).

Federal Interagency Stream Restoration Working Group (FISRWG). 1998. *Stream corridor restoration: Principles, processes, and practices.* Federal Interagency Stream Restoration Working Group (FISRWG), GPO Item No. 0120-A; SuDocs No. A 57.6/2:EN 3/PT.653. ISBN-0-934213-59-3. http://www.usda.gov/technical/stream_ restoration/ (accessed August 24, 2007).

Maestri, B., M. E. Dorman, and J. Hartigan. 1988. *Managing pollution from highway stormwater runoff.* Transportation Research Record 1166, Issues in Environmental Analysis. Washington, DC: Transportation Research Board.

People's Daily (English edition). 2002. *China to accelerate its urbanization pace.* Available at: http://english.peopledaily.com.cn/200212/06/eng20021206_108064.shtml (accessed August 24, 2007).

Prince George's County. 2002. *Bioretention manual.* http://www.co.pg.md.us/Government/AgencyIndex/DER/ESD/Bioretention/bioretention.asp (accessed August 24, 2007).

Schueler, T. 1994. The importance of imperviousness. Center for Watershed Protection, Columbia, MD., *Watershed Protection Techniques* 1(3):100–111.

U.S. Environmental Protection Agency (USEPA). 1983. *Final report: Nationwide urban runoff program.* Washington, DC: U.S. Environmental Protection Agency.

U.S. Environmental Protection Agency (USEPA). 2000a. *Low-impact development (LID)—a literature review.* EPA 841-B-00-005. Washington, DC: USEPA. http://www.lowimpactdevelopment.org/ftp/LID_litreview.pdf (accessed August 24, 2007).

U.S. Environmental Protection Agency (USEPA). 2000b. *Principles for the ecological restoration of aquatic resources.* Rep. No. EPA841-F-00-003. Washington, DC: USEPA Office of Water. http://www.epa.gov/owow/wetlands/restore/principles.htm (accessed August 24, 2007).

Virginia Department of Conservation and Recreation (VADCR). 1999. *Virginia stormwater management handbook.* http://www.dcr.virginia.gov/soil_&_water/stormwat.shtml (accessed August 24, 2007).

Yu, S. L., and R. J. Kaighn. 1995. *Testing of roadside vegetation.* Vol. 2 of *The control of pollution in highway runoff through biofiltration.* Virginia Department of Transportation, Report No. FHWA/VA-95-R29. Richmond: Virginia Department of Transportation.

Yu, S. L., and J. Wu. 2001. Laboratory testing of a mixed-media bioretention cell. Report to Americast, Inc. Department of Civil Engineering, University of Virginia, Charlottesville, VA.

Yu, S. L., and X. Zhang. 2001. The critical storm approach for TMDL development. Paper presented at the Annual Meeting of the Water Environment Federation, Chicago, IL.

Index

T - #0372 - 071024 - C16 - 234/156/15 - PB - 9780367388003 - Gloss Lamination